LAB MANUAL

TO ACCOMPANY
THE COMPLETE GUIDE
TO A+ CERTIFICATION

Michael W. Graves

THOMSON

DELMAR LEARNING™

Australia • Canada • Mexico • Singapore • Spain • United Kingdom • United States

Lab Manual to Accompany The Complete Guide to A+ Certification

by Michael W. Graves

Vice President, Technology and Trades SBU:
Alar Elken

Editorial Director:
Sandy Clark

Senior Acquistions Editor:
Stephen Helba

Senior Channel Manager:
Dennis Williams

Senior Development Editor:
Michelle Ruelos Cannistraci

Marketing Director:
Dave Garza

Marketing Coordinator:
Stacey Wiktorek

Production Director:
Mary Ellen Black

Production Manager:
Andrew Crouth

Art/Design Coordinator:
Francis Hogan

Senior Editorial Assistant:
Dawn Daugherty

Library of Congress Card Number: 2005925898

Lab Manual to Accompany The Complete Guide to A+ Certification /
Michael W. Graves

ISBN: 1418005673

NOTICE TO THE READER

CONTENTS AT A GLANCE

TABLE OF CONTENTS

PART 1: A+ GUIDE TO PC HARDWARE MAINTENANCE AND REPAIR

PART 2: A+ GUIDE TO PC OPERATING SYSTEMS

PART 1

A+ GUIDE TO PC HARDWARE
MAINTENANCE AND REPAIR

INTRODUCTION

Welcome to the *Lab Manual to Accompany The Complete Guide to A+ Certification*. Although this manual was designed to work as a companion to the textbook by the same name, instructors should have no problem adapting this book for use with any good book on PC hardware or software. Where specific chapters are referenced in this manual, refer your students to the appropriate chapter in the book that you have chosen. Obviously, I suggest that you use *The Complete Guide to A+ Certification* as your primary text as well. It not only serves well as a textbook but will be a good reference manual down the road.

The labs included in Part 1 have been designed specifically to provide the student with some hands-on experience in the field of PC hardware troubleshooting and repairing. As such, it is rather necessary that operating systems (OS) and drivers receive more than just a passing nod. Therefore, some exercises involve the installation of an OS and the subsequent introduction of new hardware to the system after the OS has been installed.

After this section, I provide a bare minimum list of materials that will be needed to accomplish these labs. But before I do so, I would like to make a suggestion. To truly accomplish the goals of this course, it is best if the students are provided with a kit that includes the necessary tools and components required to assemble a working computer. The costs of these materials can either be added on as a lab fee or built into the tuition costs, depending on how your particular organization manages its fee structure.

If the costs of this approach are prohibitive, at the very least an older working PC-compatible computer must be provided for each student. A number of online clearance houses and auctions can be used for this purpose. A list of these services, which was accurate at the time of this writing, is provided in Appendix A. Many of these sources will also be useful if you choose to take the kit route.

Should you choose to have the students assemble their systems from scratch, I suggest that you avoid using motherboards that have too many of the components built in. It is best if they go through the process of installing these devices on their own. Please note that Labs 14 and 15 in Part 1 should be introduced early in the semester. Therefore, a student kit should include the following, at the minimum:

- Enclosure
- Motherboard
- Processor
- At least 64MB of RAM
- AGP video card
- Sound card
- Modem
- Network interface card

- Floppy disk drive

- Hard disk drive

- CD-ROM drive

- Keyboard

- Mouse

- Speakers

Students will need monitors, but they can be borrowed from other areas and do not have to be included as part of the kit. Other items that will be required as the labs progress include the following:

- Ten RJ-45 connectors per student

- A box of floppy disks

- A classroom computer with an Internet connection

- A copy of Windows 98 for each student

- A floppy disk cable

- A fully configured PC with monitor

- A network cable crimping tool

- A parallel M/F cable

- A parallel/Centronics cable

- A patch panel

- A pencil

- A punchdown tool

- A SCSI cable

- A serial M/F cable

- A USB device cable

- A USB extension cable

- An antistatic wrist strap

- An IDE cable

- Bulk Cat5 or Cat5e cable

- Either a switch or a hub with sufficient ports for each student's computer plus that of the instructor

- Several sheets of paper

- Computer toolkit

The classroom should be set up in such a way that each student will have sufficient space in which to work. In one lab, students need space to completely disassemble and subsequently reassemble a computer system. Therefore, a lot of parts are going to be strewn around.

These labs are going to provide students an opportunity to get their hands a bit dirty as they encounter a large number of the CompTIA objectives for the Core Exam. Unfortunately, because the exam is theoretical as well as practical, some objectives won't be covered. For those, you'll need to rely on your textbook. Objectives that are covered include the following:

1.1   Identify the names, purpose, and characteristics of system modules. Recognize these modules by sight or definition.

1.2   Identify basic procedures for adding and removing field-replaceable modules for desktop systems. Given a replacement scenario, choose the appropriate sequences.

1.3   Identify basic procedures for adding and removing field-replaceable modules for portable systems. Given a replacement scenario, choose the appropriate sequences.

1.4   Identify typical IRQs, DMAs, and I/O addresses and procedures for altering these settings when installing and configuring devices. Choose the appropriate installation or configuration steps in a given scenario.

1.5   Identify the names, purposes, and performance characteristics of standardized/common peripheral ports, associated cabling, and connectors. Recognize ports, cabling, and connectors by sight.

1.6   Identify proper procedures for installing and configuring common IDE devices. Choose the appropriate installation or configuration sequences in given scenarios. Recognize the associated cables.

1.8   Identify proper procedures for installing and configuring common peripheral devices. Choose the appropriate installation or configuration sequences in given scenarios.

2.1   Recognize common problems associated with each module and their symptoms, and identify steps to isolate and troubleshoot the problems. Given a problem situation, interpret the symptoms and infer the most likely cause.

4.1   Distinguish among the popular CPU chips in terms of their basic characteristics.

4.2   Identify the types of RAM (Random Access Memory), their form factors and operational characteristics. Determine banking and speed requirements under given scenarios.

4.3   Identify the most popular types of motherboards, their components, and their architecture (bus structures).

4.4   Identify the purpose of CMOS (Complementary Metal-Oxide Semiconductor) memory, what it contains, and how and when to change its parameters. Given a scenario involving CMOS, choose the appropriate course of action.

6.1   Identify the common types of network cables, their characteristics, and connectors.

You will notice as you go through the labs that the same objectives will appear in different labs, again and again. This shouldn't come as much of a surprise because each of these objectives includes a number of subcomponents. So don't get tied up in the details to the point that you don't have fun and *learn!*

WORKING WITH BINARY AND HEXADECIMAL

One of the things about working with computers is that they don't speak plain English. They only speak binary. As a technician, you need to be able to perform simple binary-to-decimal and decimal-to-binary conversions. It won't hurt if you're conversant in hexadecimal either. Therefore, in this lab, you're going to work with both of these concepts. To complete this lab, all you will need is

- A pencil

- Several sheets of paper

- A good night's sleep

Here you will be introduced to the following CompTIA exam objective:

1.4 Identify typical IRQs, DMAs, and I/O addresses and procedures for altering these settings when installing devices. Choose the appropriate installation or configuration steps in a given scenario.

EXERCISE 1: BINARY

Because binary is the basic language of the computer, all information that passes through the system consists of a series of 0s and 1s. You need to figure out how to deal with these simple values in a way that allows computing to take place. In the next few exercises, you'll see how numbers are handled.

EXERCISE 1A: CONVERTING BINARY NUMBERS TO DECIMAL

Prior to attempting this exercise, let's review how a byte is made up. If you recall from Chapter One of the text, each byte has eight bits. Within each byte is a *least significant bit* and a *most significant bit* (see **Figure L1.1**).

Therefore, it should be relatively easy to convert a binary value into decimal. Simply count out the total number of bits in the value, moving from right to left. Each bit will represent a positional value. Starting with one, double the value of each position as you move from right to left. For example, in the binary value 1 0 0 1 1 1 1 0, the values would be as follows:

$$\underline{128}\ \underline{0}\ \underline{0}\ \underline{16}\ \ \underline{8}\ \underline{4}\ \underline{2}\ \underline{0}$$

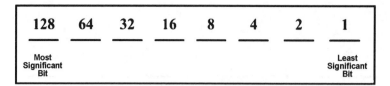

Figure L1.1   The Relative Value of Bits in a Byte

Add the numbers together, and you should arrive at 158. The preceding example is a simple 8-bit value, but the concept holds true whether you're calculating an 8-bit value or a 256-bit value. In the latter, you simply have a lot more figuring to do. With this in mind, try to convert a few binary numbers.

EXERCISE 1A REVIEW

1.   1110 0010

2.   1111 1011

3.   0000 1111

4.   0011 1001

5.   1101 0001

EXERCISE 1B: CONVERTING DECIMAL NUMBERS TO BINARY

Okay, that was easy. Now let's go in the other direction. You have a decimal number, and you want to see what it would look like in binary. You're basically going to take the procedures you used in the previous exercise and reverse them. Now, however, you're going to have to do a bit of subtraction as well. To keep things simple, stick to using 8-bit values.

If you have a number such as 148, how would you convert it to binary? First, let's look back at **Figure L1.1** and review the relative values of bits in a byte. To make things simple, you should draw your own figure of eight lines, representing each bit, and assign each line a value.

Now, look at the number you're calculating. Can you subtract 128 from 148? My calculator says it's possible, so do so and place a 1 in the 128 position, leaving a remainder of 20. So now it looks like this:

I don't care whose calculator you use. You can't subtract either 64 or 32 from 20, so put a 0 into the placeholder for each of those values. Now it looks like this:

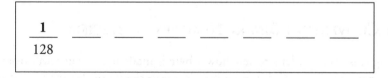

You can subtract 16 from 20, so do so, placing a 1 in the placeholder for the value of 16 and leaving a remainder of 4.

1	0	0	1				
128	64	32	16				

The rest should be easy. You have a value of 4 remaining and a placeholder for the value 4. Put a 1 in that placeholder and 0s in all remaining placeholders.

1	0	0	1	0	1	0	0
128	64	32	16	8	4	2	1

Therefore, 1 0 0 1 0 1 0 0 is the binary representation of the number 148. Pretty easy, isn't it?

One of the key places you see a decimal alliteration of a binary number is in 32-bit TCP/IP addresses. The address 192.168.0.1 is nothing more than a 32-bit value divided into four 8-bit sections and then presented in decimal form. Therefore, you'll covert a couple of TCP/IP addresses for your next few exercises.

Exercise 1B Review

Convert the following addresses to binary:

1.   141.26.215.32

2.   215.87.34.1

3.   164.111.13.136

4.   192.168.15.110

Exercise 2: Working with Hexadecimal

Of course, we mere mortals have a bit of difficulty working with binary numbers, and for certain uses, decimal don't work very well either. That's where hexadecimal notation comes into play. Binary is a Base2 notation. Decimal is Base10. If you take a byte apart, you have eight bits, which, as you should recall from Chapter One, provides for 256 total values. You really don't want to create a form of notation based on 256, now do you? Nobody has that many fingers and toes. If we break that byte down into two four-bit chunks, each of those pieces could be represented in a Base16 notation. But what do you call those numbers after you reach nine? You'll just borrow a few letters from the alphabet is what you'll do. As a result, hexadecimal counts from 0 to 9 and then from A to F (see **Table L1.2**).

With the preceding chart, you can easily convert a decimal value to hexadecimal or a hexadecimal value to binary. For example, the number 48 is easy: 48 ÷ 16 = 3. Therefore, the number 48 is represented by the hexadecimal value of 0FFFh. Typically, leading 0s are dropped, so it would actually be presented

Table L1.1   Decimal, Hexadecimal and Binary Notation

Dec	Hex	Bin	Dec	Hex	Bin
0	0	0000	8	8	1000
1	1	0001	9	9	1001
2	2	0010	10	A	1010
3	3	0011	11	B	1011
4	4	0100	12	C	1100
5	5	0101	13	D	1101
6	6	0110	14	E	1110
7	7	0111	16	F	1111

as FFFh. The lowercase h indicates that the reader is being presented with a hexadecimal value. So, with that in mind, try a few examples.

EXERCISE 2 REVIEW

■ Convert 022Eh to binary.

■ Convert the number 116 to hexadecimal.

■ Take the 32-bit value 1111 0111 0011 0001 and convert it to hexadecimal.

EXERCISE 3: WORKING WITH OCTAL

One mathematical notation used in the computer industry that is rarely introduced to technicians is *octal* notation. With octal notation, only 8 characters are used: 0 through 7. As such, three binary bits represent each character used in octal notation (see **Table L1.3**).

The reason hardware and network technicians don't see octal used very much is that it is almost exclusively the domain of the programmer. I introduce it here because I feel that anyone involved in the computer industry should be familiar with its existence, how it is used, and where it might be used. To get a feel for how octal works, you're going to learn why programmers celebrate Halloween on Christmas. (This is an old programmers' joke that goes way back. I wish I knew to whom I should give credit for its original use, but it seems several dozen people want to take credit. Because I'm not running for President, I won't claim to have invented it.)

1.   First calculate the dates for Halloween and Christmas. If you're using the same calendar I am, you should arrive at Oct. 31 and Dec. 25.

2.   Convert the number 31 in octal, which would be designated *oct* 31, to its binary value. You should arrive at 011 001.

3.   Convert 011 001 from binary to decimal. The value you obtain should be 25. In order to communicate the fact that the new value is decimal, its proper alliteration will be *dec* 25.

4.   You have now proven that Oct. 31 is the same thing as Dec. 25. Now you have an excuse for either giving gifts on Halloween or going to your company Christmas Party dressed up as a witch. (What? You already *do*?)

Table L1.2   The Hexadecimal Chart

Dec	Hex	Dec	Hex	Dec	Hex	Dec	Hex	Dec	Hex	Dec	Hex	Dec	Hex	Dec	Hex
0	0	32	20	64	40	96	60	128	80	160	a0	192	c0	224	e0
1	1	33	21	65	41	97	61	129	81	161	a1	193	c1	225	e1
2	2	34	22	66	42	98	62	130	82	162	a2	194	c2	226	e2
3	3	35	23	67	43	99	63	131	83	163	a3	195	c3	227	e3
4	4	36	24	68	44	100	64	132	84	164	a4	196	c4	228	e4
5	5	37	25	69	45	101	65	133	85	165	a5	197	c5	229	e5
6	6	38	26	70	46	102	66	134	86	166	a6	198	c6	230	e6
7	7	39	27	71	47	103	67	135	87	167	a7	199	c7	231	e7
8	8	40	28	72	48	104	68	136	88	168	a8	200	c8	232	e8
9	9	41	29	73	49	105	69	137	89	169	a9	201	c9	233	e9
10	a	42	2a	74	4a	106	6a	138	8a	170	aa	202	ca	234	ea
11	b	43	2b	75	4b	107	6b	139	8b	171	ab	203	cb	235	eb
12	c	44	2c	76	4c	108	6c	140	8c	172	ac	204	cc	236	ec
13	d	45	2d	77	4d	109	6d	141	8d	173	ad	205	cd	237	ed
14	e	46	2e	78	4e	110	6e	142	8e	174	ae	206	ce	238	ee
15	f	47	2f	79	4f	111	6f	143	8f	175	af	207	cf	239	ef
16	10	48	30	80	50	112	70	144	90	176	b0	208	d0	240	f0
17	11	49	31	81	51	113	71	145	91	177	b1	209	d1	241	f1
18	12	50	32	82	52	114	72	146	92	178	b2	210	d2	242	f2
19	13	51	33	83	53	115	73	147	93	179	b3	211	d3	243	f3
20	14	52	34	84	54	116	74	148	94	180	b4	212	d4	244	f4
21	15	53	35	85	55	117	75	149	95	181	b5	213	d5	245	f5
22	16	54	36	86	56	118	76	150	96	182	b6	214	d6	246	f6
23	17	55	37	87	57	119	77	151	97	183	b7	215	d7	247	f7
24	18	56	38	88	58	120	78	152	98	184	b8	216	d8	248	f8
25	19	57	39	89	59	121	79	153	99	185	b9	217	d9	249	f9
26	1a	58	3a	90	5a	122	7a	154	9a	186	ba	218	da	250	fa
27	1b	59	3b	91	5b	123	7b	155	9b	187	bb	219	db	251	fb
28	1c	60	3c	92	5c	124	7c	156	9c	188	bc	220	dc	252	fc
29	1d	61	3d	93	5d	125	7d	157	9d	189	bd	221	dd	253	fd
30	1e	62	3e	94	5e	126	7e	158	9e	190	be	222	de	254	fe
31	1f	63	3f	95	5f	127	7f	159	9f	191	bf	223	df	255	ff

Hexadecimal Conversion Table. Note that all leading 0's are dropped.

Table L1.3 Octal Notation

Oct	0	1	2	3	4	5	6	7
Bin	000	001	010	011	100	101	110	111
Dec	0	1	2	3	4	5	6	7

LAB REVIEW

1. 22Fh represents a _____.

    a. 12-bit hexadecimal value

    b. A 4-bit binary value

    c. A 16-bit hexadecimal value

    d. A 32-bit hexadecimal value

2. You are receiving repeated error messages that say, "A fatal exception has occurred at 1EEF80CE:1EEE0110." This is most likely _____.

    a. An I/O address

    b. A memory address

    c. A bad memory chip

    d. A CPU failure

3. A TCP/IP address is an example of _____.

    a. A hexadecimal alliteration of a binary value

    b. A decimal alliteration of a hexadecimal value

    c. A binary alliteration of a decimal value

    d. A decimal alliteration of a binary value

4. Describe two different places where a hexadecimal value might be used to troubleshoot an issue.

5. TCP/IP addresses in IPv6 are going to be represented by _____.

    a. Decimal values

    b. Binary values

    c. Hexadecimal values

    d. A combination of decimal and binary values

LAB SUMMARY

Understanding binary and hexadecimal needs to be second nature to the technician. You don't need to be so fluent in either notation that you can glance briefly at a value and spout off its decimal equivalent (although such an ability would make you the envy of all your peers). You do need to be able to recognize a binary or a hex value when you see one, and you need to be able to differentiate between 16-bit, 32-bit, and 64-bit hex values. These are the types of values that frequently show up in error messages or Event View messages. You can use them to track down the source and/or meaning of a given event.

LAB 2

IDENTIFYING PORTS AND CABLES

In this job, the ability to look at a plug and know right where it goes must be second nature. Not only does your state of mind improve but imagine that of your client. If you don't even know where to plug the keyboard, how is she supposed to have confidence in the job you did rebuilding her entire computer? Therefore, in this lab, you will make a closer examination of each of the ports, both on paper and up close. To complete this lab, you will need the following:

- A fully configured PC with monitor
- A parallel M/F cable
- A parallel/Centronics cable
- An IDE cable
- A floppy disk cable
- A SCSI cable
- A serial M/F cable
- A USB device cable
- A USB extension cable
- A pencil and paper
- A partner

The following CompTIA exam objective will be covered in these exercises:

1.5 Identify the names, purposes, and performance characteristics of standardized/common peripheral ports, associated cabling, and their connectors. Recognize ports, cabling, and connectors by sight.

PART ONE: THE PORTS

This section is divided into two exercises. One will be on paper and should be completed individually. The second involves the computer, and it is best if everyone takes on a partner. By the time you're finished, you should be comfortable with all the common ports on a computer system.

13

EXERCISE 1A

In the blank beneath each of the following illustrations, or on a separate sheet of paper, identify the port being shown.

Figure L2.1_____

Figure L2.2_____

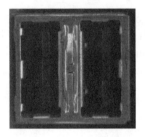

Figure L2.3_____

Figure L2.4_____

Figure L2.5_____

Figure L2.6_____

Figure L2.7_____

Figure L2.8_____

EXERCISE 1B

This next exercise is a little more complicated. You're going to do basically the same thing as you just did, only you're going to identify the parts as they reside on a physical computer. And, you're going to do it without looking! Working with a partner, position yourself in front of a computer system. Then reach around with your hands, and by the sense of touch alone, identify the following components:

- VGA port
- Serial port
- Parallel port
- Power connector
- Modem port
- NIC interface
- USB port
- Sound card

Your partner will let you know whether you're right or wrong on each guess. When you are finished, switch places, and you can pick on your partner when he or she makes a mistake.

EXERCISE 1B REVIEW

1. Which port had nine pins?
2. Which port had fifteen pins?
3. How could you differentiate between the modem and the NIC?

PART TWO: THE CABLES

The one exercise in this section is a group effort. Your instructor will pass around examples of several different cables so that you can examine them closely and know what they look like when you encounter them in the field. The cables distributed should include the following:

- 40-conductor IDE
- 80-conductor IDE
- Parallel cable
- Printer cable
- Serial cable (25-pin and 9-pin if both are available)
- SCSI cable (as many different types as the instructor can assemble)
- USB device cable
- USB extension cable

After these cables have been distributed and discussed, each student should remove the monitor, keyboard, and mouse cables from the back of the classroom machine and examine them. Note that the keyboard and mouse cables are likely to be identical, so you might want to note the position they were in before you removed them. If you forget, just remember that on a tower case, the keyboard cable generally is positioned closer to the edge of the case.

PART TWO REVIEW

1. How was the physical connector on the 80-conductor IDE different than that of the 40-conductor cable?

2. How was the printer cable different from the parallel cable?

LAB REVIEW

1. Name two different connectors that each has 15 pins. How can you differentiate between them?

2. Name three different connectors with 25 pins each and describe the difference. (*Hint:* For the third, you'll need to pour through the text.)

3. Why is it such a good idea to be able to identify ports and connectors on the back of a computer system by touch alone?

4. What is the only 15-pin connector with three rows of pins used on PCs?

5. What one feature of most modems makes it easy to differentiate the modem from the NIC simply by feel?

LAB SUMMARY

Recognizing the basic components of a computer system has to be a second nature to any professional technician. It just doesn't look professional for the Field Service Representative for a major manufacturer to hold up a component in front of the customer and say, "I wonder what the heck THIS is!" For some reason, that sort of behavior fails to establish the proper rapport between the technician and the customer. Cables are just as important. How would you like the guy fixing your furnace to turn to you and ask, "I don't suppose you remember what I pulled this off of, do you?"

A Tour of the Inner Realm

In Lab Two, you got a feel for what's shaking on the outside of the computer. Now, I'm going to have you dive right in and take a closer look at what makes up the system. You're even going to take out the motherboard and put it back in when you're finished. To do this, you'll need the following:

- A working computer for each student
- Your toolkit
- An antistatic wrist strap
- A pencil and paper

In this lab, you're going to get some hands-on experience with a couple of different CompTIA exam objectives. Among them are the following:

1.1 Identify the names, purpose, and characteristics of system modules. Recognize these modules by sight or definition.

1.2 Identify basic procedures for adding and removing field-replaceable modules for desktop systems.

1.3 Identify basic procedures for adding and removing field-replaceable modules for portable systems.

4.3 Identify the most popular type of motherboards, their components, and their architecture (bus structures).

EXERCISE 1: PROCEDURES FOR OPENING A COMPUTER

Wouldn't it be marvelous if there was but a single set of instructions for this procedure that would apply to every enclosure? Alas, the variety of enclosures on the market is exceeded only by the variety of insect life in the Amazon River basin. There are as many ways to open a computer enclosure as there are flavors of ice cream. Therefore, I'm going to list some of the more common approaches. In each example, however, step one is always to disconnect the power cord and all other cables from the back of the case. Make *sure* that you're wearing your antistatic wrist strap properly affixed to ground.

- *The Removable Side/Top Panel:* This enclosure is probably the most common, so I'll describe it first. With this configuration, the enclosure frame of a tower has separate right and left side panels that are easily removed. On the desktop case, it is the top panel that can be removed. On the tower, as you face the front of the enclosure, the panel you want to remove is on your left. Lay the system on its right side. The method of detaching this panel varies somewhat, but it is rarely difficult to figure it out. Two or more screws might be holding the panel to the back of the frame; there might be a hinged release; or release buttons might be on either side of the rear edge of the panel. After you remove the screws or press the release mechanism, the panel will either slide toward the rear, or you will need to lift it upward. In some rare instances, the panel slides toward the front.

- *The U-Cover:* Although generally seen on older enclosures, a couple of manufacturers still espouse this design. With this design, four or more screws hold the cover onto the frame. When you remove these correctly, the left, right, and top panels all come off in a single piece.

- *The Hinged Enclosure:* This enclosure was made popular by Dell but subsequently has appeared on other models. The reason for its popularity is how fast and easy it is to access all parts in the system. Two releases hold the top section of the enclosure in place. When released, the case opens up like a giant book. The drive bays and front panel circuits are all on the top/front section, which is now out of the way of the bottom section that holds the power supply, motherboard, and associated components.

EXERCISE 1 REVIEW

1. Based on your reading here and in the text, describe some advantages of the hinged enclosure over the U-cover.

2. If you have Internet access, do a search for computer enclosure manufacturers and browse the offerings available.

EXERCISE 2: REMOVING THE MOTHERBOARD

As with enclosures, manufacturers have come up with a wide variety of methods for securing the system board to the case. Once again, I'll describe the most commonly encountered methods. However, in nearly all cases, some preliminary procedures must be performed before removing the system board.

1. *Remove all expansion cards.* With the vast majority of designs, a single screw holds the expansion card to the back plane. Some designs have a plastic or metal clamp that comes down over the back planes of all the expansion cards at the same time. Release a clip and lift the clamp. Be gentle removing the cards. You don't want to damage either the circuit board or the bus connectors.

2. *Remove all cables from the system board.* At a minimum, you will encounter the power cable from the power supply to the system board, a ribbon cable attached to each drive (a single cable may be attached to multiple drives), and connectors from the front panel. Front panel wires might be clustered into a single connector, or a number of different, and very small, connectors might be marked with labels such as HDD, LED, PWR, and so on. If the system board you're working on is of the latter type, take your pencil and paper and diagram where each of these wires attach. No universal standard for how these wires hook up exists, and it is very unlikely that you have access to the system board manual. Also, Pentium 4 and equivalent machines have a second power connector coming from the power supply to the motherboard.

3. *Remove the memory modules.* With most modern systems, two clips are on either end of the module. Press them downward, and the memory module will pop out of the socket. If you have

an older machine that uses SIMMs, two clips are located on either end of the module, but toward the front of the socket. Release these, and the module will tip forward about 15 degrees and you can lift it out. Place the memory on an antistatic surface.

4. *If the CPU is a socketed CPU, remove the heatsink/fan assembly.* Generally, a retaining clip is on either side of the heatsink that holds it in place. They need to be pressed down and outward at the same time. Some of the better designs have a slot into which the tip of a flat-bladed screw-driver can be fitted. If the assembly you're working on adapts only to fingers, keep your language to yourself. (Others are present.) If it is a slotted CPU, the CPU and heatsink/fan come out as a single assembly. Two vertical clips on either side of the CPU hold it in place. Release the clip and lift the CPU out. Make sure that the fan is unplugged from the motherboard. On the socketed CPUs there is a lever on one side of the CPU socket. Press downward and to the side, and then lift up. You can now lift the CPU out of the socket.

5. *This step might or might not be required, depending on the model.* You'll know. With some enclosures, one or more drive bays must be removed before the system board can be lifted from the enclosure. With some cases, a single screw attaches the lower drive bay to the upper drive bay. On some designs, two screws attach the lower drive bay to the front of the enclosure. In this case, you'll have to remove the front panel. Once in a while, you'll encounter a design in which the lower drive bay simply rotates out of the way.

6. Another step that you might need to perform on some systems but not others is that of removing the power supply. In many enclosures, the power supply simply tilts out of the way on a hinged bracket, so this is unnecessary. On others, it isn't in the way to begin with. If it does need to be removed, look for the following methods of mounting the power supply:

   • The most common method is four screws on the back panel of the enclosure attaching the power supply to the frame. Remove those four screws and disconnect all power supply leads from the motherboard and drives. Lift the power supply out of the enclosure and set it aside.

   • Another method seen on the so-called "tool-free" enclosures is a spring-release mechanism that unlocks the power supply from the frame. This is usually clearly marked and sometimes even painted a bright color to make it stand out. Unfortunately, I've also seen enclosures on which the release button blended in with the rest of the enclosure so well that I had to call Tech Support to figure out how to remove the power supply.

7. *Now it's time to remove the system board.* With most designs, several screws (usually five to eight) attach the motherboard to the frame. Remove the screws, slide the board a half inch or so toward the front to free the I/O ports from the rear of case, and lift the board out. Another design to look for is a release clamp holding the motherboard in place. This clamp is usually located toward the front of the enclosure. A tray that holds the motherboard slides out, and you lift the board out, tray and all.

EXERCISE 2 REVIEW

1. What is a key precaution to take prior to beginning the removal of a motherboard?

2. Why is it advisable to remove RAM before taking the motherboard out of the case?

EXERCISE 3: A TOUR OF THE MOTHERBOARD

With the rest of the computer pushed aside, place the motherboard on the table and look for each of the following components. Your instructor will assist you by pointing them out.

- CPU socket
- Chipset
- Memory sockets
- BIOS chip
- Battery
- IDE ports
- Floppy disk port
- AGP slot
- PCI slot
- ISA slot (if present)
- Power connector
- Wiring harness
- Power connector for CPU fan

EXERCISE 3 REVIEW

1. How can you differentiate between the PCI slot and the AGP slot?

2. What is an additional connector you'll find on a Pentium IV motherboard that won't appear on the Pentium III or Celeron boards?

EXERCISE 4: THE PAPER MOTHERBOARD

From **Figure L3.1**, identify each of the labeled components. This time, you're working without the help of your instructor. Whether or not you work with a partner is the instructor's choice.

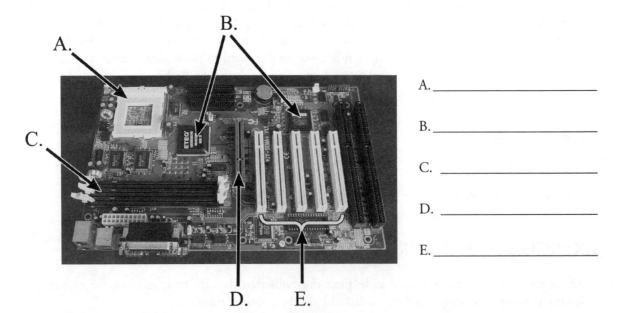

A. _____

B. _____

C. _____

D. _____

E. _____

Figure L3.1   Identify the parts to which the arrows point.

LAB REVIEW

1. List and describe as many different methods of opening a system enclosure as you can.

2. What is the correct order for removing the following three components when pulling a motherboard from a system: screws, cables, and expansion cards?

3. What are some different ways in which a motherboard might be attached to the enclosure frame?

4. How many different types of cables are you likely to find inside of a typical computer system? Describe them.

5. List as many different motherboard components as you can.

LAB SUMMARY

Just as it is critical for the professional technician to know his or her way around the outside components of a computer system, the parts on the inside are just as important. Maybe even more so. What makes the inner realm a little more complicated is that far fewer standards govern how the inside of a computer system is designed. Therefore, the technician has to get used to a bit of variety and a few surprises now and again. I've been doing this since the days of the 286, and I *still* come across situations in which I find myself muttering, "Where the heck did they hide the power connector for the CPU fan?"

INSTALLING IDE DEVICES

As a technician, one of the most common tasks you'll find yourself doing is that of installing a new drive. You're always adding or replacing hard drives, CD burners, DVD drives, and so on and so forth. For the most part, it's a fairly straightforward procedure. But you'll occasionally encounter a few pitfalls. Knowledge is the key to avoiding those pitfalls. In this lab, you'll need the following:

- The lab PC
- An IDE hard disk drive
- An IDE optical drive (CD, CD-RW, or DVD)
- A 2-channel IDE cable (preferably an older style 40-conductor, non-cable select version)
- The toolkit

Over the course of this lab, you're going to get up close and personal with a couple of CompTIA's exam objectives as well. These objectives include the following:

1.2 Identify basic procedures for adding and removing field-replaceable modules for desktop systems.

1.6 Identify proper procedures for installing and configuring common IDE devices.

1.8 Identify proper procedures for installing and configuring common peripheral devices.

2.1 Recognize common problems associated with each module and their symptoms, and identify steps to isolate and troubleshoot the problems.

EXERCISE 1: REPLACING A HARD DRIVE

When replacing a hard drive, one of the first things you need to do is take out the old one. As I pointed out in Lab Three, different manufacturers have different methods by which they mount hard drives in a system. So once again, I'll point out the most common. All of these methods involve some form of dedicated drive bay in which the drive resides.

- *Standard Drive Bays.* The vast majority of designs involve a standard drive bay. When you open the case, you will see where the external drives are installed. Directly below these are two or more

3.5-inch drive bays. They might or might not be removable. One is external for the floppy disk drive. One or more additional externally accessible bays might exist for adding tape drives or Zip drives. The ones you're interested in are the ones with no access to the outside world. If the drive bays are removable, there will be either a screw affixing the cage to the 5.25-inch external bays above it or one or two screws affixing it to the front of the frame, or it may simply rotate upward out of the case.

- *Front Panel Drive Bays.* These are more commonly seen with micro-ATX mini-tower cases. The drive mounts vertically in a bay located on the inner surface of the front panel of the frame. Four screws hold the drive in place. To access these screws, you must remove the front plastic (check with the specific manufacturer on how that is done).

- *Power Supply Drive Bays.* These aren't seen very often any more because of the heat generated by today's higher-rpm drives, but they were fairly common in older designs. One or more 3.5-inch drive bays were located directly beneath the power supply, and in some cases, one of the drives actually was affixed to the power supply.

The preceding list doesn't represent every possibility, but it certainly represents the vast majority of machines you're likely to encounter today. After you've figured out which design you have, you can proceed to the removal process.

1. Remove the IDE and Molex cables from the drive. In smaller enclosures, it's best to remove them completely from the system and set them aside. They can get in the way. If the cables aren't keyed, make a note of which direction the #1 conductor on the cable is pointed. That's the wire on the ribbon cable colored pink or red. If you get it backwards later on, you won't do any damage, but the drives on that cable sure won't work!

2. If you're working with a removable drive bay, remove it. If not, make sure that both side panels are removed from the enclosure.

3. Remove the four (usually, anyway) screws affixing the drive to the cage. As you can see in **Figure L4.1** and **Figure L4.2**, these screws may be located on the sides of the drive or on the base. All hard drives are designed to be mounted either way.

4. Slide the drive out of the bay.

Now that wasn't so hard, was it? I'd like to say that replacing the new drive is simply a reversal of the preceding process. And for the most part, it is. But in the replacement procedure, you need to add a step.

Figure L4.1 In this illustration, you can clearly see the screw holes on the sides of the drive used for mounting it into the system. You also see the 40-pin connector, the power socket, and the master/ slave jumpers if you look closely enough (and reproduction holds up).

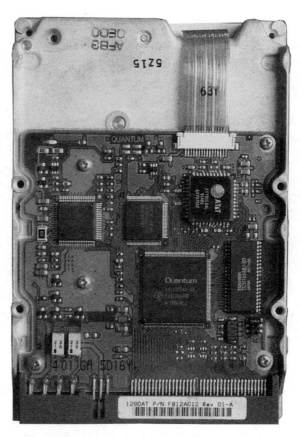

Figure L4.2 In this illustration, you can see the base-mounting screw holes. You also see the controller circuitry for the drive. Most modern drives have this circuitry enclosed. I chose this drive for illustration purposes simply because it did not.

Before installing the new drive into the system, make sure that the master/slave jumpers are set properly. It's pretty frustrating to spend five to fifteen minutes fighting with a poorly designed micro-ATX enclosure mounting the drives only to find out that your new drive isn't being recognized by the system. Also, it's a good idea to attach the IDE cables to the drive while it's still outside of the system and you have plenty of room to work.

TIP: When working with removable drive bays, it's a good idea to attach the IDE cables before you put the drives back into the case. You've got more room in which to work. Also, if you're working with 80-conductor cables—which is most likely to be the case on any system using newer drives—you have a couple of other things to consider. 80-conductor cables like the one in **Figure L4.3** are cable-select by default. The master drive will go on the end connector, and the slave goes onto the middle one. You want to put the master drive in the top bay and the slave beneath it if you're installing two drives. Otherwise, you'll end up having to twist your cable around to make it work properly. Over time, ribbon cables that have been twisted to extreme angles can fail due to metal fatigue in the 24-awg wires used in their design.

As you're sliding the drive into the bay, before it's all the way back into the enclosure and while you still have some room in which to manipulate your fingers, now is the time to attach the Molex connectors. In the vast majority of enclosures, when the drive is in place, you have one or two inches between the back of the drive and the power supply. If your hands are as big as mine, or if you suffer from Coordination Deficiency Syndrome (CDS) as I do, then trying to attach the Molex after the drive is in place can result in Tricky Terminology that can't be taught in this class.

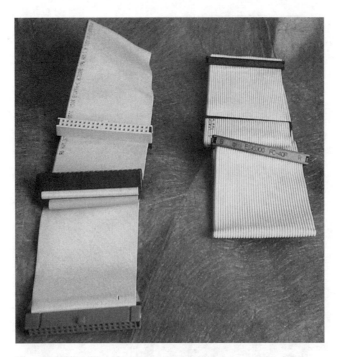

Figure L4.3 The 80-conductor cable seen on the left is cable select by default. 40-conductor cables, like the one on the right, might or might not be cable select. If a notch has been clipped from the section between the middle and end conductors on the 28th wire, you know it is. However, if the manufacturer simply didn't attach that wire to the connector, you can't tell just by looking.

Don't just replace the drive back in your machine however. Pull the shunt off the jumper. (Don't drop it! Manufacturers somehow know what color of carpet you're working over and always make the shunt the same color.) Then swap drives with another classmate. Now you can install your new drive, making sure that it is properly jumpered for your system. If you're using the older style non-cable select 40-conductor cables like I suggested in the introduction to this lab, you're going to have to figure out which setting to use.

EXERCISE 1 REVIEW

1. You've just opened up the enclosure and there are no apparent internal drive bays for a hard disk. Where else might the manufacturer have hidden the drive?

2. What additional step do you commonly have to perform in order to remove hard disks that have been attached to the front of the frame?

3. What is the name of the cable commonly used to provide power to the hard drive?

4. What are the two locations for screw holes on 3.5-inch hard drives?

5. What are the three possible jumper settings for an IDE drive?

EXERCISE 2

In this exercise, you're going to make two different IDE devices coexist on the same IDE port using a 2-device cable. To make it more of a challenge, the class will be divided into two sections. You'll see why in a few minutes.

1. Open your computer and remove the hard drive and the CD-ROM. Also pull the IDE cable(s). If the systems used in your class had each device on a separate single-channel cable, set aside these cables and substitute a cable designed to support two devices.

2. Pull the shunts from the jumpers of both drives.

3. Take both drives and their shunts over to a person in the other section and trade with him or her.

4. Install the new drives into your system, after making sure both drives are jumpered properly.

After you've installed the drives and confirmed that they are both working properly, put your systems back together. You're finished with this lab.

EXERCISE 2 REVIEW

1. What are two things that can cause one IDE device to be seen on one channel, but not the other?

2. If neither IDE device can be seen, and both are set to cable select, what is the first thing you might want to check?

SOME IDE TROUBLESHOOTING TIPS

For the most part, IDE drives are fairly trouble-free in both installation and configuration. The advent of cable-select drives and cables makes it a simple matter of setting the jumper on each drive to cable-select and making sure that your primary disk is on the end connector. A few things might get in the way of this working to your benefit. Among these are the following:

■ Some drives need to be master drives on the channel. This might include primary boot drives on older systems and some models of CD-RW. If these drives are configured to be a slave, they'll either not be recognized by the system or not function properly after the system has booted.

■ If the device that needs to be the slave on the cable is positioned too far away for you to attach the middle connector on the cable to that device and then loop the end connector down to the primary device, then you might have to put these devices on separate channels. A master device *cannot* be attached to the slave connector on a cable-select cable.

■ If the IDE channel being used has been disabled in the CMOS, then you can tear your hair out trying to figure out why you can't get a device to be recognized by the system. With many versions of BIOS, instead of telling you that the channel has been disabled, you simply see the word NONE for devices on that channel. Change the setting to AUTO.

■ Make sure to attach the Molex connector to the drive. Amazing as it may sound, a hard drive can be compared to your refrigerator because if it doesn't have an available supply of electricity, it won't work.

■ Do not reverse the IDE cable on either the drive or the motherboard connector. Not all IDE cables are keyed, and if your cable is plugged in backward, the drive won't be recognized. You won't hurt anything, but it won't work.

LAB REVIEW

1. You've just installed two drives on the same cable, and now one or both of the drives can't be recognized by the system. How many things can you think of that might cause this?

2. Why would having one drive set to be cable select and the other drive set to be a slave cause problems?

3. How many drives can be installed on a single IDE port?

4. What are two differences between 80-conductor IDE cables and 40-conductor cables, aside from the obvious difference in the number of wires? How are they similar?

5. If an IDE cable is plugged in backward on either the drive or the port, what is the extent of the damage that will be done?

LAB SUMMARY

Now that you've had some hands-on experience fighting with IDE devices, they won't even cause you to break a sweat when you're in the field. But as you all know, if there's a possibility of something going wrong, it will. Sometimes what looks like a bad hard drive is actually a bad cable, or even a bad motherboard. After you've replaced your drive and it still doesn't work, do the following:

■ *Double-check your jumpers.* You don't want one drive set to cable select and the other to slave.

■ *Check your cable.* If you have a cable-select cable and the drive on the end is set to be a slave, it'll never be seen by the system. The drive at the end must be the master.

■ *Make sure that all cables are properly seated and positioned properly in the socket.* With those larger ribbon cable connectors, it's pretty easy to have one side not pushed all the way in. Also, not all IDE cables are keyed. When they're not, they are very easy to plug in backwards.

■ *Make sure that the power cables are attached to the drives.* I can't count the number of times I've been summoned by a student who said that the drive wasn't being recognized by the system, only to discover that the power cable wasn't plugged in. (Of course, *I've* never done that!)

HARD DISK PREPARATION

Simply installing a hard disk into the computer system doesn't make it suddenly start working. The master boot record, which I described in the textbook, and the file allocation tables have to be generated before it is useful. For that, you're going to need some specialized software and some time. For the next couple of exercises, you'll need the following:

■ The computer into which you installed the hard disk in Lab Four hooked up to a keyboard and monitor. A pointing device will not be required (or useful).

■ A bootable floppy disk with Microsoft's FDISK and FORMAT utilities (available at www.mwgraves.com in zip format)

In the course of these exercises, you will be exposed to the following CompTIA exam objectives:

1.2   Identify basic procedures for adding and removing field-replaceable modules for desktop systems.

1.3   Identify basic procedures for adding and removing field-replaceable modules for portable systems.

1.6   Identify proper procedures for installing and configuring common IDE devices. Choose the appropriate installation or configuration sequences in given scenarios. Recognize the associated cables.

EXERCISE 1: AN OVERVIEW OF FDISK

Microsoft operating systems (OS), including Windows 2000 and later, no longer include FDISK as their disk preparation utility. Disk preparation is a part of the installation process for the OS. This is good in that it has made installing an OS easier for novices. It is bad in that, should you need to prepare a disk for use outside of that operating system, you can't work from a floppy disk. Most technicians I know still hang on dearly to a floppy disk made from either MS-DOS or Windows 98 (or earlier).

A usable disk can be made from a Windows 98 machine. Put a blank, formatted floppy disk into the drive and then click Start → Settings → Control Panel. Now double-click Add/Remove Programs. In the window that opens, click the tab that says Startup Disk. With the blank disk in the drive, click the Create Disk… button. You may be prompted for the Windows 98 CD as well. If a Windows 98 machine is not available, a zip image of a boot disk is available at www.mwgraves.com.

Let's begin.

1. With the boot disk in the drive, start the machine. After it has gone through POST, it will deliver you to an A: prompt (**Figure L5.1**). For those of you who have never set foot outside of a graphical interface, try not to be intimidated, but you're going to actually have to type some commands here.

Figure L5.1 The Command Prompt

2. At the A: prompt, type **fdisk** and then press the <Enter> key. The opening screen will appear (**Figure L5.2**). All that fancy explanation is simply asking you whether you want to format your drive to FAT16 or FAT32. If you choose Y (yes) when it asks whether you want to use large disk support, you have selected FAT32. If you press N (no), you will select FAT16. If for any reason you need FAT16, you should press N. For the purposes of this (and later) exercises, press Y.

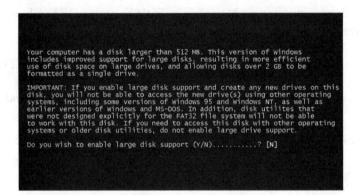

Figure L5.2 The Opening Screen of FDISK

3. The FDISK Options screen appears next (**Figure L5.3**). It is from this screen that the various tasks are selected. Which task you select depends entirely upon what you're trying to do. On a new, unprepared hard drive, you would select Option 1: Create DOS partition or Logical DOS

```
                    Microsoft Windows 95
                   Fixed Disk Setup Program
            (C)Copyright Microsoft Corp. 1983 - 1995

                        FDISK Options

Current fixed disk drive: 1

Choose one of the following:

1. Create DOS partition or Logical DOS Drive
2. Set active partition
3. Delete partition or Logical DOS Drive
4. Display partition information
5. Change current fixed disk drive

Enter choice: [1]




Press Esc to exit FDISK
```

Figure L5.3 The FDISK Options Screen

Drive. If you are working with a disk that has previously contained data, you would want to start by selecting Option 4: Display partition information. Option 5: Change current fixed disk drive, will appear only in systems that have more than one physical hard disk installed. FDISK can recognize when there is only a single hard disk and, in that case, Option 5 will not appear. For this exercise, select Option 4.

4. The screen shown in **Figure L5.4** will appear. The information contained in that screen may vary from class to class and from computer to computer, depending on what was previously installed on that hard drive. If it's a new hard drive, there should be a message telling you that no DOS partitions exist.

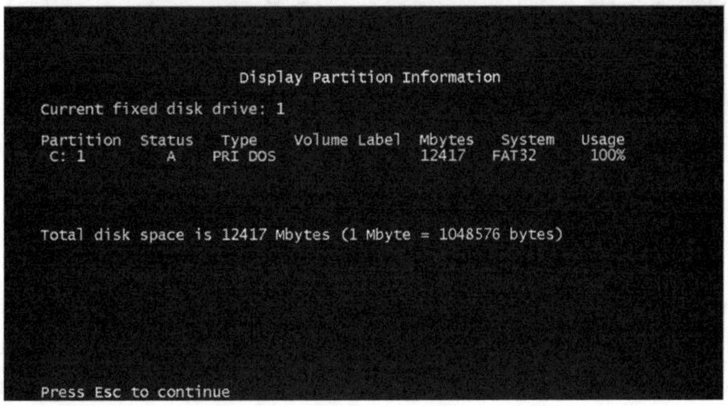

```
                         Display Partition Information
      Current fixed disk drive: 1

      Partition  Status   Type   Volume Label   Mbytes   System   Usage
         C: 1      A     PRI DOS                  12417   FAT32    100%



      Total disk space is 12417 Mbytes (1 Mbyte = 1048576 bytes)






      Press Esc to continue
```

Figure L5.4 Viewing Partitions in FDISK

5. Press the <Esc> key to return to the FDISK Options screen. If your system contained previous partitions, you need to delete them. To do so, press Option 3: Delete partition or Logical DOS Drive. This will bring up a screen like the one shown in **Figure L5.5**.

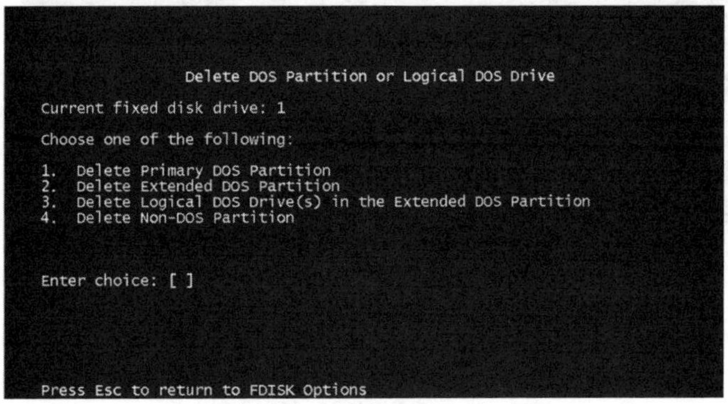

```
                   Delete DOS Partition or Logical DOS Drive
      Current fixed disk drive: 1

      Choose one of the following:

      1. Delete Primary DOS Partition
      2. Delete Extended DOS Partition
      3. Delete Logical DOS Drive(s) in the Extended DOS Partition
      4. Delete Non-DOS Partition


      Enter choice: [ ]




      Press Esc to return to FDISK Options
```

Figure L5.5 Deleting a Partition in FDISK

6. If extended partitions are on your drives, you must first select Option 3: Delete Logical DOS Drive(s) in the Extended DOS Partition and delete each logical drive on the partition. Next, you need to delete the extended partition, and finally you can proceed to Option 1: Delete Primary DOS Partition.

7. This brings up the screen shown in **Figure L5.6**. You are warned that to continue will delete all data on your drive. Select Y, and a new field will open up prompting you to enter the Volume Name. If your drive has a volume name, you must enter it exactly as it appears on the screen.

You then will be given yet another option to bail out when it asks whether you're really sure. Press Y to delete the partition, and now the previous data on the hard drive is history. This is when far too many technicians scream, "NO, WAIT! THAT WAS THE WRONG DRIVE!!!"

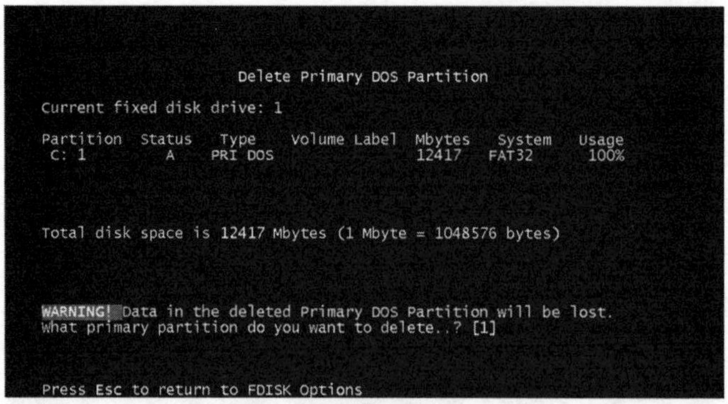

Figure L5.6   You get one last chance to bail out!

8.   Press <Esc> to return to the FDISK Options menu. Now you want to press Option 1: Create DOS partition or Logical DOS Drive. This will bring up the screen shown in **Figure L5.7**. Before you can create any extended partitions, a Primary DOS partition must first be created. Press Option 1: Create Primary DOS Partition. You will be asked whether you want to use all available space and to make the partition active. Select Y.

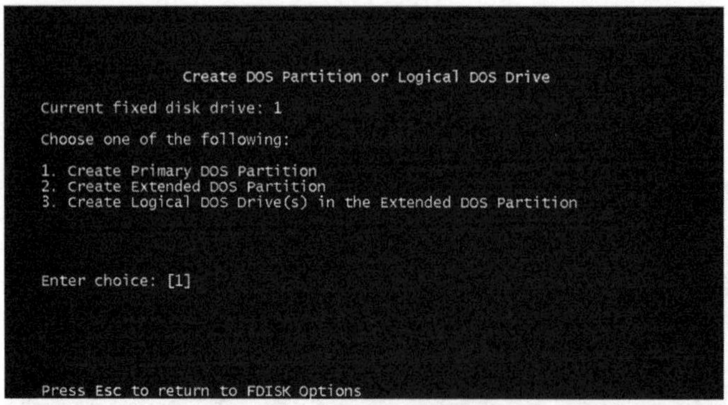

Figure L5.7   Creating a DOS Partition

9.   FDISK then will quickly scan the disk for obvious flaws, calculate available space, and create the primary partition. Press <Esc> twice to exit FDISK. You will be prompted that in order for the change to take effect, you must restart your machine. Make sure the boot disk is still in Drive A:, select Y, and let the machine reboot. You are now ready for Exercise 2.

Exercise 1 Review

1.   Why is it a good idea to view partitions on a previously existing hard disk before making any permanent changes?

2.   If you divided your hard disk into multiple partitions using FDISK, what additional step must you complete before you will have a bootable drive?

SOME COMMENTS ON FDISK

Keep in mind that FDISK is an older utility and, as such, is subject to some limitations. For example, a partition created by a non-Microsoft OS such as Linux frequently can't be recognized. When you view partitions, you see a Non-DOS Partition listed, but when you try to remove it, you are informed that no Non-DOS partitions exist. It can drive you nuts. Even Microsoft OSs occasionally do something to the MBR that prevents FDISK from being able to work with that disk. I frequently work in classroom environments where the OS of a given system changes from day to day. I discovered that in some cases, Windows 2000 Advanced Server would make the disk unrecognizable after the students had converted the drives to Dynamic Disks. This is an option in Windows 2000 that applies advanced properties to the drive that earlier OSs don't know how to deal with. Oddly enough, I never encountered that problem in classrooms using Windows 2000 Professional.

Along a different line, should you want to create multiple partitions using FDISK, you have two options for dividing your disk into partitions. In Step 8, instead of selecting Y when asked whether you want to use all available space, select N. Now you will be encouraged to enter how much of the drive you want used for the primary partition. If you type a number, such as 150, you are defining how many megabytes of space that partition should contain. A number followed by the percent (%) sign is obviously a percentage. If you type 50%, it will take exactly half of the available space and assign it to the primary partition. Either way you choose, the result of your actions will be that your primary partition is *not* set as active by default. You'll have to go back to the initial FDISK menu and select Option 2: Set Active Partition and select which of the partitions you've created should be active.

3. Using FDISK, you want to remove an extended partition, but it won't let you. What does the error message tell you that you must do before removing extended partitions?

4. You created two separate partitions on the same hard disk, using the % Disk Space method of allocating disk space to each partition. Now what must you do before you have a bootable system?

5. In the opening screen of FDISK, you selected N when asked whether you wanted to enable Large Disk Support. Just what did you do to the system when you made that selection?

EXERCISE 2: FORMATTING A HARD DISK

Partitioning the hard disk created a new master boot record, but the disk still isn't usable. You now need to generate the file allocation tables. This is done when you format the hard disk. So let's do it!

1. The first thing you need to do is determine what your new drive letter is going to be. If this is a second drive in your system, you don't want to be formatting Drive C. That's when the View Partition Information option you saw in the previous exercise comes in handy. If it is the first and/or only drive in the system, you simply type **format C:** at the command prompt.

2. As you see in **Figure L5.8**, you are warned that continuing will destroy all data on the drive. Do you wish to proceed? Well, as I was preparing this lab, I was using the same computer I'm using to write the manuscript. So I chose to press N for no. You should press Y for yes and sit back and wait. Depending on the size of the hard drive, your age, and how long it is until lunch, this can take anywhere from several minutes to the rest of your natural life. When it finishes, you're finished with this lab.

SOME COMMENTS ON FORMATTING

The format command used by DOS and Windows is known as a high-level format, or sometimes as the operating system format. A low-level format is done at the factory, which maps out the sectors and tracks on each platter. Some older versions of BIOS contain an option for Low Level Format. This is for SCSI or older MFM hard drives *only* and should *never* be used on an IDE drive. Using that utility on an IDE drive will turn it into a paperweight. Some of the better hardware diagnostic utilities include a low-level format for IDE drives. This might be useful in the event that a failed attempt to install an operating system has left Track 0, Sector 1 unreadable. It should, however, be used only as a last resort.

Also, for the initial format of a newly partitioned hard drive, it can take quite some time to format the drive. What the format utility is doing is going out to each file allocation unit (FAU) on the drive, testing its integrity, and then going back and writing the FAT entry. Any bad FAUs will be marked as such and will not be reported to the OS. Subsequent formats of that disk can be much quicker. Using the command format /q will perform a quick format that only rewrites the FAT without doing the surface scan.

Something to keep in mind when reinstalling operating systems is that it is frequently a good idea to use FDISK on the hard disk before formatting an existing drive. The reason for this is that the operating system owns the FAT. Reinstalling the same OS can result in unpredictable problems.

Figure L5.8   Formatting the Hard Disk

EXERCISE 2 REVIEW

1.   Why is the Quick Format so much quicker than a standard format?

2.   Referring back to the textbook, answer the following questions. You've just formatted the hard disk and several dozen bad sectors were found.

   a.   Why didn't the overall capacity of the drive get reduced accordingly?

   b.   Did you even know it happened?

LAB REVIEW

1. Using information provided in this lab along with information from the text, explain just what is going on during the FDISK routine when you create a primary DOS partition.

2. Using knowledge you gained in this lab as well as from the text book, explain what happened to the MBR when you divided the disk into multiple partitions.

3. Explain what is going on during the format process and just why it takes so much longer to perform an initial format than it does when you select the format /q option.

4. Using information from this lab and from the text, explain the difference between a low-level format and an operating system format.

5. *For extra credit:* Even after using FDISK and formatting the hard drive, more work must be done by the OS during its installation. What might that be?

LAB SUMMARY

Well, now you know that either replacing or installing a new hard drive isn't simply a matter of bolting in the new drive and hooking up a couple of cables. A certain degree of drive preparation needs to be done as well. It's all part of the job.

INSTALLING AN OS AND MAKING A BOOT DISK

For the remaining exercises in this book, it is essential that an Operating System (OS) be installed on each machine. Of course, in the real world, it is necessary to have an OS as well, and you can't really call yourself a technician until you've done an installation. I have chosen Windows 98 as the OS for these labs for two reasons. First, as of this writing, there are still an estimated 40,000,000 machines in the United States running Windows 98. One of these days, you're going to be called on to fix one of them. Second, the more modern Microsoft OSs make hardware installation almost transparent to the end user. I want to make you do the work yourself. For this lab, you're going to need the following:

- The lab machines, complete with keyboard, monitor, and mouse
- The boot disks you created in Lab Five
- A copy of Windows 98 or Windows 98SE for each machine
- A blank, formatted floppy disk for each student

There isn't a whole lot in this lab directly related to the CompTIA objectives, but it is still a good lab to complete. This lab has a lot of real-world application of technique. What you will see related to the exam includes the following:

1.2   Identify basic procedures for adding and removing field-replaceable modules for desktop systems.

1.3   Identify basic procedures for adding and removing field-replaceable modules for portable systems.

1.6   Identify proper procedures for installing and configuring common IDE devices. Choose the appropriate installation or configuration sequences in given scenarios. Recognize the associated cables.

2.1   Recognize common problems associated with each module and their symptoms, and identify steps to isolate and troubleshoot the problems.

EXERCISE 1: INSTALLING WINDOWS

For this exercise, each student needs to have access to a lab machine and a boot disk with CD-ROM support. If you used the Windows 98 Startup Disk utility or downloaded the boot image from www.mwgraves.com, then you have what you need. So let's install Windows. (*Note:* Windows 98 ships on a bootable CD, so if for any reason a boot disk is not available, Setup can be started from the CD as long as the machine's CMOS is configured to boot to CD-ROM. The purpose of booting to the floppy in this exercise is to point out a minor difference in the process.)

Insert the boot disk into the floppy disk drive and start the machine. If this is the first time you've ever booted to floppy, don't be concerned about the amount of time it takes. That's normal.

When you boot a computer using the WIN98 Startup Disk, a *ramdisk* will be created. A ramdisk is an area in physical RAM that is treated exactly as if it were a hard disk. It is divided into 512-byte sectors just like a hard disk, and a file allocation table is created for that disk. The Windows Utilities will be loaded to this disk for faster performance. This can throw off the drive letters from what you might be expecting, because the ramdisk will become Drive C until your newly FDISKed hard disk has been formatted. After the hard disk is formatted, the ramdisk will become Drive D, moving your CD-ROM up to Drive E.

1. When you've established the correct drive letter for your CD-ROM, type the command `d:\setup` from the A: prompt, where *d* is replaced by the actual drive letter of your CD-ROM. You will be notified that Setup is going to perform a routine check on your system. Press <Enter>.

2. Microsoft's ScanDisk utility will run an abbreviated check on all hard drives installed in your system. If it discovers any problems, the installation may be aborted. If not, it will tell you that it discovered no problems. Either way, you have either the option of viewing the log or exiting ScanDisk. Press <Tab> to move over to Exit, and then press <Enter>.

3. This brings up the Windows 98 Setup screen. Setup will give you an estimated path that your setup will take, based on the information it collected during the routine check of your system. Now you should have mouse support. Click Continue.

4. Setup begins preparing the Setup Wizard. This will take several seconds.

5. Next you will have the option of installing Windows to the C:\Windows directory or to select another directory. Unless you have a driving need to do otherwise (such as another version of Windows already installed on the system), select the default setting and click Next →.

6. The Setup Options screen appears next. Here, you're going to deviate a bit from a standard installation so that I can point out some interesting and very useful options that don't get installed if you select a Typical installation. Click Custom and then click Next →.

7. The Windows Components screen will appear. Note that the checkboxes next to the various options appear in three different ways:

   ☑ Indicates that all available options in this category have been selected.

   ☑ Indicates that some, but not all options have been selected. Click the Details button to view the options that have been selected and those that haven't been.

   ☐ None of the options have been selected. Click Details to see what options exist.

8. Select each category in turn to see what options haven't been selected by a Typical installation. In particular, select System Tools, click Details, and make sure that Backup, System Monitor, and System Resource Meter are installed. Click OK and then Next →.

9. Now you need to identify your computer for the network. Because a later lab will involve networking these machines, now is as good a time as any to make sure that your computers' names are user friendly, rather than the randomly selected collections of letters and numbers that Setup chooses for you. Here, the instructor should select a starting computer and each one should be named Student1, Student2, Student3, and so on until all systems have been given a unique but easily remembered name.

10. On the Computer Setting screen that appears, leave the defaults as they are, unless, of course, they aren't correct for your region. If not, change them accordingly and click Next →.

11. The Establishing Your Location screen appears next. Select the appropriate location and click Next →.

12. Now you will be given the opportunity to create a Startup disk. I don't want to do that right now, so click Next →. Setup will try to create a disk anyway. It will start copying files and then prompt you to insert a blank formatted floppy. It will also warn you that any information on that disk will be destroyed. Click Cancel.

13. This will bring up the screen labeled Start Copying Files. As you have most likely already figured out, this will start the file copying process. Click Next → and gather into groups to discuss the football playoffs until the process has been completed. Or you can just sit there and read the advertisements that pop up for a product you've already bought. Depending on the speed of your machine, the estimated time remaining might range anywhere from half an hour to your next birthday.

14. After the file copy process has completed, your machine will reboot automatically. If you want to save a couple of seconds, take the CD out of the drive before it reboots. If not, just select the option to boot from the hard disk on the boot menu that appears if there is a bootable CD in the drive.

15. Windows will start with the message "Starting Windows for the First Time" embedded in the splash screen. It will eventually settle onto the User Information screen. Type your name. The Organization field is optional. Click Next →.

16. The licensing agreement appears. Press Page Down as you read the entire agreement. (Make certain you read every single word.) Click I accept the Agreement to continue. If you try to be funny and decline Microsoft's generous terms, the installation will be aborted, and the rest of the class will have to wait for you until you get caught back up. Or the instructor may kick you out of class and tell you to come in on your own time to finish this lab.

17. The Windows Product Key screen will now appear. Type the 25 letters and numbers that make up the key. It doesn't matter if you use the shift key for capitals or not. Letters will be entered as capitals anyway. If you mistype any character, you'll have to try again before you can continue.

18. The Start Wizard tells you that it will now save all the information and that you should click Finish. That's why it's called the Start Wizard.

19. It now initializes the Driver Database and scans your computer for installed hardware devices. Any Plug 'n Play device for which there is a driver in the WIN98 database will be automatically installed during the next step. Those not recognized will need to be installed manually after the installation is completed. When the hardware scan is finished, your computer will restart once again.

20. The Setting Up Hardware screen will be the next thing you see. Setup is now installing the device drivers for the hardware it was able to detect.

21. The Date/Time Properties screen will now appear, and you will have the opportunity to set the computer's time and date. By default, the selected time zone is Pacific.

22. Now Setup will configure the Control Panel, the Start Menu, Windows Help, how Windows will handle DOS programs, and the configuration settings.

23. When this is finished, your computer will restart once again. As Windows starts this time, which may take a bit longer, it will attempt to install devices that weren't installed during the

previous hardware installation process. This will include devices hooked up to peripheral ports such as your monitor and non-Plug 'n Play devices. For the latter, in order to successfully install these devices you will need the device drivers on either floppy disk or CD-ROM.

24. The Welcome to Windows 98 screen will now appear. In the lower left corner of this screen, deselect the checkbox that says Sho<u>w</u> this screen each time Windows 98 starts and close the screen.

EXERCISE 1 REVIEW

1. You've just booted your computer to a Windows 98 Startup disk, and your CD-ROM drive has just been renamed from Drive D to Drive E. What happened?

2. What is the first thing the Windows Setup program does after it starts?

3. In the Windows Components screen, some of the boxes are grayed out with check marks inside. What does this indicate?

4. What happens when the first phase of file copying has been completed?

5. When the Licensing Agreement appears, you say out loud, "What a crock!" and click I Disagree. Now what happens?

EXERCISE 2: CREATING THE TECHNICIAN'S BOOT DISK

The Windows 98 Startup Disk is fine and works quite well. However, it carries a lot of excess baggage you don't need for a basic troubleshooting boot disk and doesn't include a couple of utilities that you'll find useful. In this exercise, you create the Technician's Boot Disk.

1. Open a command prompt. To do this, you can either click Start → Programs → MS-DOS Prompt, or you can click Start → Run and type **command** into the command field. If your disk needs to be formatted before using it, you can type **format a: /s** at the command prompt.

2. Insert a blank formatted high-density floppy disk into Drive A and type **sys a:** at the command prompt. Your computer will think about it for a few moments and then copy some files to the floppy and create an MBR on that floppy. The disk is now bootable, but it isn't finished to our satisfaction.

3. To open Windows Explorer, right-click the Start button and select Explore. Browse to the C:\Windows\Command directory, right-click the file mscdex.exe, and select copy. Now right-click A: and select Paste.

4. Browse to the C:\Windows directory and locate the files emm386.exe and himem.sys. Copy those files to your floppy.

5. Now, insert the original boot disk that you used to start your original installation. You're looking for one of two files. I've seen both used from time to time. The file you're looking for on the floppy is either mtmcdai.sys or oakcdrom.sys. Right-click this file, select copy, and then paste it to the C:\Windows\Temp directory. Put the disk for your new boot disk back in the drive and cut and paste the file from C:\Windows\Temp to A:\.

6. Back at the command prompt, type **edit**. The MS-DOS Editor will appear. First, you want to create a config.sys file. Type the following lines. Press the <Enter> key after each line.

```
FILES=32

BUFFERS=32

Stacks=9,256

LASTDRIVE=Z

DEVICE=HIMEM.SYS

DEVICE=EMM386.exe

DOS=HIGH

DEVICEHIGH=MTMCDAI.SYS  /D:CDROM
```

7. From the Editor menu, select File → Save As, and in the file name field, type **config.sys**.

8. Now, select File → New, and you'll create an autoexec.bat file. In the Editor screen, type the following lines.

```
PROMPT $P$G

MSCDEX.EXE  /D:CDROM
```

9. Save this file, and you're finished with this part. But there's still more work to do on this disk.

10. From the C:\Windows\Command directory, copy the files edit.com, fdisk.exe, format.com, xcopy.exe, xcopy32.exe, and xcopy32.mod.

11. To test your new disk, leave it in the drive and reboot your machine. When you're finished, save the disk. You'll be using it later to modify the Technician's Boot Disk.

EXERCISE 2 REVIEW

1. What are two different ways to open a command prompt in Windows 98?

2. What is the correct command for making a formatted disk bootable?

3. You have a disk that needs to be formatted before you can use it to create a boot disk. What command can you use to format the disk and make it bootable at the same time?

4. What files will you copy from the C:\Windows directory of a Windows 98 machine to your Technician's Boot Disk?

5. What are the two files that you need to create on your boot disks?

LAB SUMMARY

Okay, you now have a workable OS on your computer. For many of you this wasn't the first time you've done this. But for those of you who have never installed an OS before, it should have been a satisfying experience. Windows 98 is a bit more cumbersome than some of the more recent Microsoft offerings, but it is a lot easier than other OSs such as Novell or Linux. When you take the Operating Systems section of the A+ series, you'll see what I mean.

WORKING WITH CONTROL PANEL

Technically speaking, the operating system is not part of this class. However, since configuring hardware to work with a computer includes getting the OS to recognize that hardware, I find it essential to take two short excursions into the OS. In this lab, I shall take you on a tour of the Windows Control Panel. In the next lab, I'll introduce you to the Registry Editor. The only thing you'll need for this lab is the computer onto which you installed Windows 98 in the previous lab, equipped with a monitor, keyboard, and mouse. No core exam objectives are covered in these exercises.

EXERCISE 1: AN OVERVIEW OF THE CONTROL PANEL

Everything that makes Windows do what it does is controlled by the Registry. But the Registry can be a pretty scary place to be, even for the seasoned veteran. Therefore, Microsoft has provided the Control Panel for adjusting the Registry settings that are most commonly manipulated. You can get to the Control Panel in two ways. One is to double-click My Computer and then double-click Control Panel. The other is to click Start → Settings → Control Panel. Either way, you wind up with a screen like the one in **Figure L7.1**.

As you can see, this screen has a number of icons. These are shortcuts to the various applets of the Control Panel. As other services are installed onto the OS, other icons are added as well. For example, in the figure, there is an icon for HP JetAdmin. You will most likely not have that icon. I purchased a Hewlett Packard Jetdirect as a more efficient method of sharing the printers on my network than simply hooking up a printer to somebody's computer and sharing it out. When I installed the software, the service was added to Control Panel. Also the icon with the inverted V labeled Modem Settings was added when I installed a USB modem. Other icons you're not likely to see on your screen include Administrative Tools, Adobe Gamma, and Automatic Updates. For the most part, you should see the rest of the icons in Figure L7.1 on your screen. Now let's go over what each applet controls.

ACCESSIBILITY OPTIONS

Accessibility Options (**Figure L7.2**) enables you to customize the way your keyboard, display, and mouse function. Although these options are primarily designed to assist the disabled as they use a computer, many of these features are useful to people without disabilities. The various features that can be configured here include the following:

Figure L7.1   The Control Panel

- *StickyKeys:* Enables simultaneous keystrokes while pressing one key at a time
- *FilterKeys:* Adjusts the response of your keyboard
- *ToggleKeys:* Emits sounds when certain locking keys are pressed
- *SoundSentry:* Provides visual warnings for system sounds
- *ShowSounds:* Instructs programs to display captions for program speech and sounds
- *High Contrast:* Improves screen contrast with alternative colors and font sizes
- *MouseKeys:* Enables the keyboard to perform mouse functions
- *SerialKeys:* Allows the use of alternative input devices instead of a keyboard and mouse

ADD/REMOVE HARDWARE

Double-clicking this icon starts the Add/Remove Hardware Wizard (**Figure L7.3**). For the most part, Windows can automatically detect and configure any new Plug 'n Play hardware you install. Should you decide to install something that is not Plug 'n Play, or should Windows fail to install a device, the Add/Remove Hardware Wizard comes in handy. It will do a scan of your system and detect any devices that are installed. It can also detect devices that were installed but have not been properly configured to run in Windows. The Add/Remove Hardware Wizard first tries to auto-configure devices. If it fails, it leads the user through a step-by-step process of properly installing that device.

ADD/REMOVE PROGRAMS

Back in the days when Windows 98 was considered state of the art, this applet was a lot more useful than it really is today. Still, even today, it's a very useful application. Or, at least it would be if more people actually used it. It isn't the adding of programs that make it useful, but rather the removal of said programs (**Figure L7.4**). However, in the days of Windows 98, in order for the applet to properly remove a program, that program needed to be installed by the applet as well. When you use Add/Remove programs to install

Figure L7.2   Accessibility Options

Figure L7.3   Add/Remove Hardware

a program, Windows maintains a log of everything that is changed. New folders that are created and new Registry entries that are added get recorded in this log. When you decide to uninstall that program later, Add/Remove Programs uses the uninstall log to back off all changes that were made during installation. Without that log, the majority of new folders and Registry entries remain on the system.

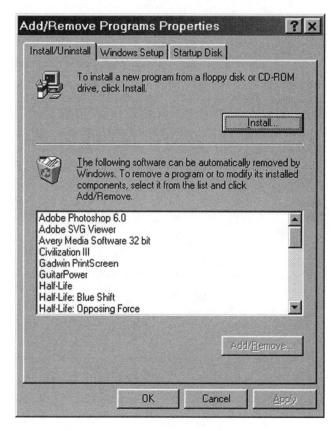

Figure L7.4   Add/Remove Programs Properties

DATE AND TIME

Hopefully, this applet doesn't require a great deal of explanation. It's, um, where you set your date and time (**Figure L7.5**). However, a third function is located in this applet that you might find useful if you ever move across the country—the ability to change the time zone in which you reside.

Figure L7.5   Date and Time Properties

FONTS

This is another often overlooked applet in the Control Panel. For most of us, our font collection grows as a result of installing new software. Any new application that provides a collection of fonts assumes that you want to use everything it has to offer and installs every single one into your system. The problem with this shotgun approach is that the more fonts you install, the longer it takes your system to load at startup. Then when you decide to use something other than your normal typeface, you have to scroll through hundreds of choices. In the Fonts applet (**Figure L7.6**) you can install and/or uninstall fonts to create the list useful to you. Of course, once again, like most of us, you have no idea what some of these fonts look like.

You have two ways to sort out the wheat from the chaff. To view a font, double-click its icon. You'll get a typeface sheet that shows the font in several different sizes. You can print that sheet if you so desire. Another useful trick this applet does is to sort fonts by similarity. Click View → List Fonts by Similarity. When you highlight any given font in the list, other fonts are listed as being Very Similar, Fairly Similar, or Not Similar. From what I can tell with these feeble eyes, Very Similar means identical. If you have two dozen fonts that all show up as being Very Similar, you're better off picking one and uninstalling the rest.

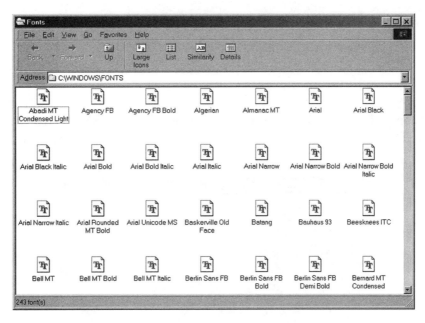

Figure L7.6   Fonts

INTERNET PROPERTIES

Internet Properties (**Figure L7.7**) is a very busy applet and one with which the average technician spends an inordinate amount of time. This is where you configure all the settings that let your computer successfully communicate with your Internet Service Provider (ISP). But it doesn't stop there. It's also where you can go to make sure that your children aren't hopping onto Websites like www.crazedmilitants.com and other sites that you would rather them not be seeing. It is in this applet that you configure how many days worth of history to maintain and to empty the history at your convenience. To go into every setting to be found in this applet is beyond the scope of this lab, but you might want to spend a few extra moments poking around to see what there is to see.

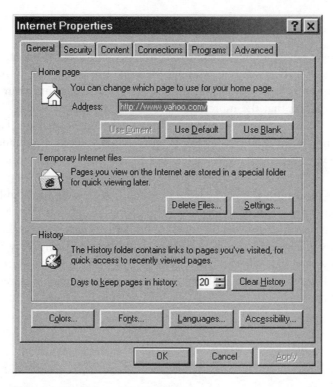

Figure L7.7 Internet Properties

KEYBOARD

The Keyboard applet (**Figure L7.8**) is, coincidentally enough, the place where certain keyboard settings are configured. These settings include the kind of keyboard installed, the repeat rate for keys that

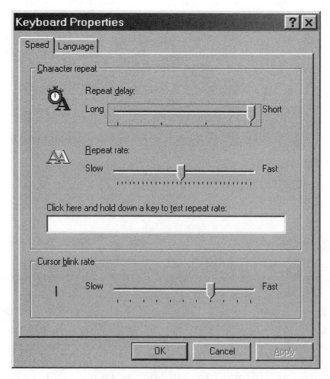

Figure L7.8 The Keyboard Applet

automatically repeat the character as long as the key is depressed, and how fast the cursor blinks. In case there's anybody out there who really cares about some of those things.

MOUSE

How many times have you sat down at somebody else's computer and found out that the mouse was set up to be left-handed? Or that the click rate was so slow that you entered the next millennium between clicks? Or so fast that Superman couldn't move his fingers fast enough to get the desired results? The Mouse applet (**Figure L7.9**) is where you change all those settings and more. In this applet, you can also change the mouse cursor, enable mouse trails (not on *my* computer, you don't!), and configure how much movement of the mouse is required in order to move the cursor a certain distance. (Drive your boss nuts. Make the cursor move as fast as it possibly can, set the click rate to maximum speed, and then enable mouse trails. Then delete the Mouse applet from Control Panel.)

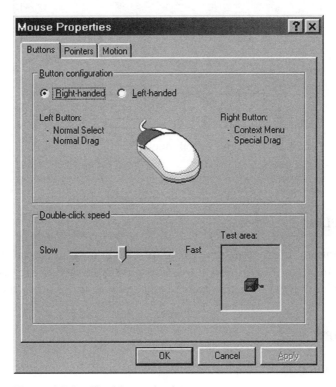

Figure L7.9 The Mouse Applet

NETWORK

Another applet that you'll spend a lot of time exploring is the Network and Dial-up Connections applet (**Figure L7.10**)—especially if you pursue a career as a network administrator. This is where all local area connections for the local area network are configured in Windows 98. In Windows 2000 and XP, dial-up connections were added to this applet.

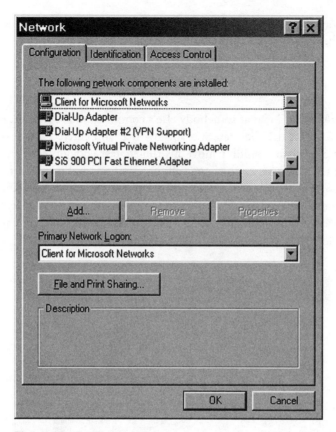

Figure L7.10   The Network Applet

THE SYSTEM APPLET

To the hardware technician, the System applet (**Figure L7.11**) is far and away the most useful applet within Control Panel. In fact, it is so important that Exercise 2 is devoted to a review of that applet alone, which means that you're finished with this exercise and ready to move on to the next.

EXERCISE 1 REVIEW

1. In terms of function, what is the purpose of Control Panel?

2. How does Add/Remove Programs make uninstalling software easier, assuming that you used the utility when you installed the program to begin with?

3. Which tab in the Internet applet enables you to configure which URL as your preferred home page?

4. What are two key functions of the System applet?

5. Why would two different computers display two different collections of utilities in Control Panel?

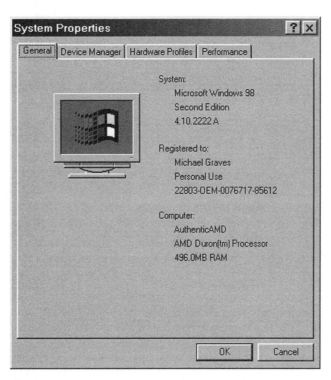

Figure L7.11 The Systems Properties Screen

EXERCISE 2: AN OVERVIEW OF THE SYSTEM APPLET

When you double-click the System icon, the System Properties window that you saw in **Figure L7.11** appears. Four tabs are at the top of that screen. These tabs are General, Device Manager, Hardware Profiles, and Performance. By default, the Systems Properties screen opens to the General page. You can change nothing on this screen, but it does show some good information. It tells you what version and build of Windows 98 you're using. It tells you to whom the product is registered and records the CD key that was used to install the software. It tells you how much memory is available to the system, and it tells you the make and model of CPU running. Note the oddball amount of memory running on this particular system. This is because this particular system has on-board video. There is 512MB of RAM installed, but 16MB have been allocated to video. Because that's not available to the system, Windows reports 496MB as being available. Now let's look at the other screens.

1. *Click Device Manager.* The screen shown in **Figure L7.12** appears. I deleted the driver files from the NIC and the sound card so that you would have some icons other than the normal ones to view. The yellow exclamation point you see next to PCI Ethernet Controller is the result of that. It tells me that it found a device installed in the system but couldn't find the driver for that device. A red X across the icon would tell me that the Registry tells Windows that a device driver has been installed for a particular device, but Windows doesn't detect the device. Hopefully, your systems won't have any warnings like this.

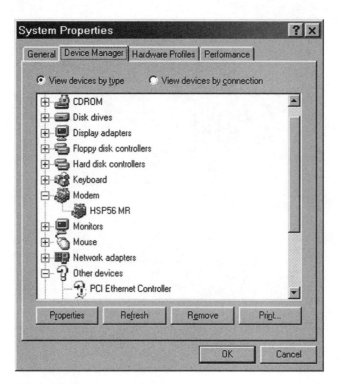

Figure L7.12   Device Manager

2. With Computer highlighted, click the Properties button. This will bring up a screen like the one on **Figure L7.13**. Here, you can view the system properties by IRQ, I/O address, or DMA channel or view how system memory is being used. By default, Interrupt request (IRQ) appears first. If you scroll down this screen, you'll see how the 15 available IRQs have been allocated in your system. You will also note that the same IRQ is used more than once. How this works is covered in Chapter Ten, Examining the Expansion Bus, of the textbook.

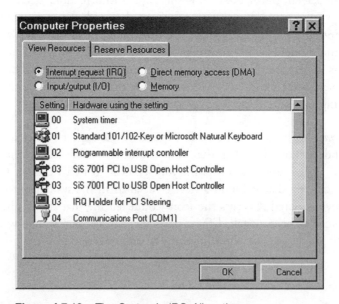

Figure L7.13   The System's IRQ Allocations

3. *Now click Input/output (I/O).* You should get the screen shown in **Figure L7.14**. This shows the base I/O address for every device installed on the system along with areas of buffer memory that have been assigned to that device by Windows. Any time you get repeated memory errors that always occur at the same address, check this list to see whether that address falls within one of the ranges listed here. If so, you know what device is causing the problem.

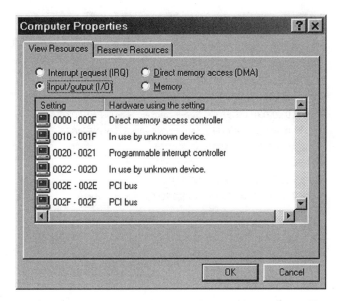

Figure L7.14   I/O Properties

4. *Next, take a look at Direct memory access (DMA).* As **Figure L7.15** shows, there isn't a whole lot to see here. Very few devices use system-arbitrated DMA these days. About the only thing you would see in this list is the floppy drive, a sound card, and an LPT port if it is configured to ECP.

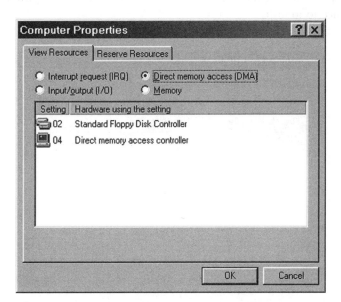

Figure L7.15   DMA Properties

5. *Finally, examine the Memory page (Figure L7.16).* What you see here can vary greatly from machine to machine. If your lab consists of several different makes and models of computer system, don't expect all of them to have the same information here. The numbers you see here are the hexadecimal addresses of the device drivers for the devices listed alongside the address. As with I/O addresses, these can be useful in tracing those so-called memory errors Windows is always reporting.

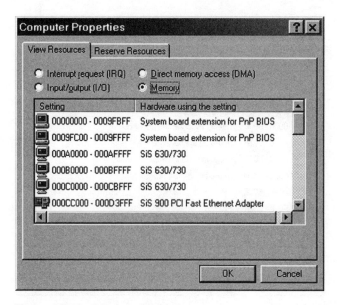

Figure L7.16 Memory Properties

6. Now for that Reserve Resources screen I've been ignoring so far. If you click that tab on any one of the properties screens you just examined, you get a blank screen like the one in **Figure L7.17.** If you click the Add button, you will be prompted to enter a value relative to the type of resource you're trying to reserve. By doing this, you are effectively taking that resource away from Plug 'n Play, and it will not be allocated to any device. This is useful if you have a device that can be configured only manually, and Plug 'n Play keeps stealing the resource before your device has a chance to grab it.

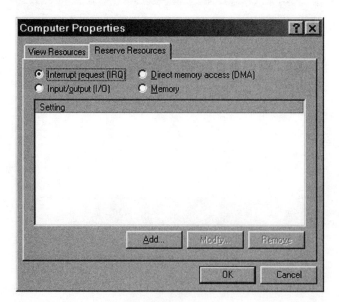

Figure L7.17 Reserve Resources

7. *Click Cancel to get back to the Device Manager screen.* Then click the Hardware Profiles tab. You should get the screen shown in **Figure L7.18**. Notice that, so far anyway, you only have one profile listed. It is named Original Configuration.

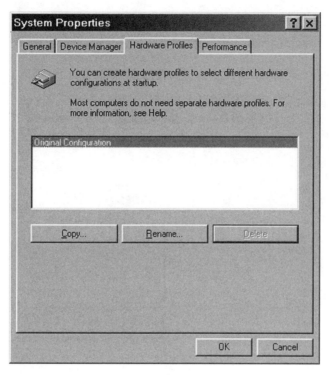

Figure L7.18   Hardware Profiles

8. *Click Copy, and in the new screen that pops up, type **New Profile***. Now go back to Device Manager and highlight one of the devices listed there. Preferably pick a device that might not be required every time you boot the machine. I have selected the modem in **Figure L7.19**. If you look toward the bottom of the Properties screen for that modem, you'll see a checkbox labeled Disable in this hardware profile. Check that box, as I have done.

9. *Click OK.* You will see the red X I described earlier in this lab. Click Close and reboot your machine. When POST has completed, as Windows is starting, you should get a boot menu offering you the following options:

   a. Original Configuration

   b. New Profile

   c. None of the above

10. *Select New Profile and let the boot process continue.* If you try to use the device you disabled, you'll be in for an unhappy surprise.

11. Finally, I will introduce you to the options found in the Performance tab of the System applet. Click that tab to get the screen shown in **Figure L7.20**.

12. *Click File System to get the screen in **Figure L7.21**.* I'm not going to take the time to go through each and every one of the options listed here. But, if you like, take a look at each of the options listed to see what you could change.

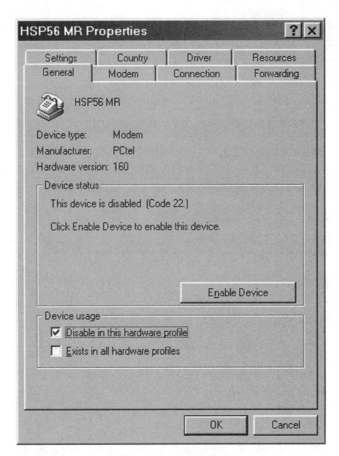

Figure L7.19 Device Properties

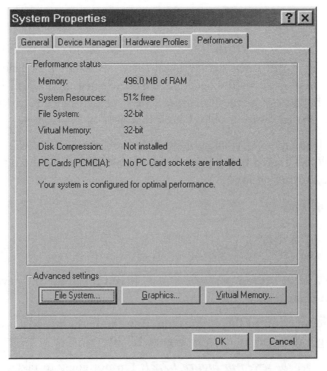

Figure L7.20 Adjusting System Performance in the
System Applet

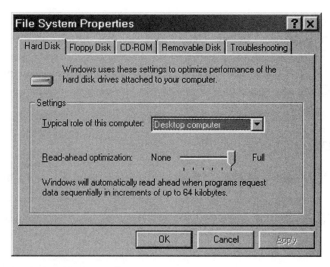

Figure L7.21   File System Properties

13. Pressing the <u>G</u>raphics button on the Performance tab provides only one option: how much hardware acceleration you want to apply. Unless your system is giving you fits, the default setting of Maximum should work fine.

14. Clicking the <u>V</u>irtual Memory button on the Performance tab brings up the screen shown in **Figure L7.22**. There is rarely any need to change from the default setting of Let <u>W</u>indows manage my virtual memory settings (Recommended). This setting adjusts the size of your swap file in Windows. Your swap file is an area of hard disk space that is treated as if it were installed memory. If your swap file is too small, system performance will drop dramatically. Therefore, if your hard disk is filling up, you might want to configure a fixed swap file so that you'll run out of disk space before your swap file is cut down to unacceptable levels.

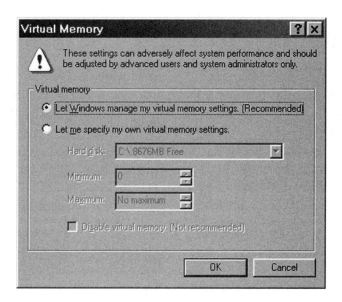

Figure L7.22   Setting Virtual Memory

15. As you can see, Control Panel is a very handy program. Therefore, the last thing you're going to do is create a shortcut for it on the Desktop. To do so, double-click the My Computer icon on your desktop. Right-click the Control Panel icon and select Create Shortcut. A message will pop up warning you that you can't create a shortcut here and do you want it on the desktop instead? Click OK, and the shortcut will appear on the desktop instead.

EXERCISE 2 REVIEW

1. What are four different views of the Computer Properties that you can bring up in Device Manager?

2. What would be the purpose of reserving specific resources in Device Manager?

3. What would be the effect of reserving those resources on the system?

4. What are two reasons you can think of for creating two or more different hardware profiles on a system?

5. What would be one good reason for manually configuring the size of the swap file in Windows?

LAB REVIEW

1. Just what is the purpose of Control Panel, and what settings are you manipulating when you make changes there?

2. Define two different ways to access the Control Panel.

3. What is one reason why Accessibility Options might not appear in your Control Panel?

4. How would you create a shortcut for Control Panel on your desktop?

5. What are four different sets of resources that can be viewed on the Computer Properties screen?

LAB SUMMARY

Control Panel is a pretty complex space, isn't it? I might not have covered some of the icons that appear on your screen, or I might have covered some that didn't appear; it's all in the way you chose to install Windows. As I pointed out earlier, adding new services will add new icons. Get to know Control Panel as well as you can, even if you're not personally a Windows user by nature. Ninety percent of your customers will be.

AN OVERVIEW OF THE REGISTRY EDITOR

As I pointed out in Lab Seven, the Control Panel is a safe place for making changes to the Registry for the most common issues. But once in a while something comes up in Windows that requires that the Registry be surgically edited. Should this need arise, *be careful!* With some of the Registry settings, having a single character missing or out of place can render your system unstable or even unbootable. Fortunately, Windows users have access to a utility that enables them to back off to a previous version of the Registry. This utility varies from version to version. As with Lab Seven, all you need to complete this lab is the student computer equipped with keyboard, monitor, and mouse. Once again, no specific Core Exam objectives are covered here. This is simply information you need to survive.

> CAUTION: **Be very careful when editing the Registry. An incorrect setting can make a device not work properly or can even cause the system to become unbootahble. Before making any changes, make sure you back up the Registry using the instructions found in this lab.**

EXERCISE 1: AN OVERVIEW OF THE REGISTRY EDITOR

1. The Registry Editor is one of those utilities that Microsoft has deliberately concealed from the average user. There are no pretty little icons to click on. It's a command line utility that can be run either from a DOS window or from the Start → Run command line. In Windows 98, this utility is launched by typing the command regedit at either the command prompt or from the Start → Run command line. Windows 2000 and XP users have two different versions of this utility at their disposal. Typing **regedit** gets basically the same utility that Windows 98 users have. A safer, but perhaps less potent version, comes from typing **regedt32** at the prompt. The latter does not have quite as powerful a search function, but it does allow the user to set security settings on individual Registry keys within the editor. Since regedit is available to both, this is the one I'll review.

2. Click Start → Run and type **regedit** in the command line. You'll get the window shown in **Figure L8.1**. The left-hand window of the screen is known as the *navigation area,* and the right-hand window is the *topic area.*

3. The Windows Registry has six different *keys* that are stored collectively in two different files. The files are user.dat and system.dat. A third file is created optionally when System Policies are enabled. This file is called config.pol. The keys are as follows:

Figure L8.1 The Opening Screen to the Registry Editor

a. HKEY_CLASSES_ROOT: This key contains the information needed for linking objects between different applications, determining what file types are opened by what applications and for mapping specific functions to keystroke patterns or mouse clicks.

b. HKEY_CURRENT_USER: Here is where user-specific information is stored. This would include items such as the programs in the user's start menu, desktop settings, applications that appear on the desktop, and display preferences.

c. HKEY_LOCAL_MACHINE: This is where information specific to the computer is stored. Device drivers, installed software, and information specific to installed hardware can be found in this key.

d. HKEY_USERS: Windows supports the ability to allow several different users to log on to the same machine, and if so desired, each user can have their own specific settings and preferences. This key stores the preferences and settings for all users.

e. HKEY_CURRENT_CONFIG: As you saw in Lab Seven, it is possible to set up multiple hardware profiles on the same computer. Whereas HKEY_LOCAL_MACHINE stores all hardware and software information, HKEY_CURRENT_CONFIG loads the information specific to the profile chosen during boot.

f. HKEY_DYN_DATA: This key stores dynamically configured information concerning the status of Plug 'n Play at the time of boot. Changes to device settings that do not require a reboot are managed by this key. This key is created on the fly by Windows at each startup and is not stored in any permanent file.

4. Beneath each of these primary keys are collections of *hives*. Hives are subkeys that contain information specific to a particular aspect of the machine. Click the + next HKEY_LOCAL_MACHINE in the navigation area to open up the hive. You should get a screen similar to **Figure L8.2**.

5. If all you did was click the + sign, as instructed, there is still nothing in the topic area. Open the Software hive and then the Microsoft folder and scroll all the way down to Windows. Open the Windows folder and highlight CurrentVersion. You should get a screen similar to that seen in **Figure L8.3**.

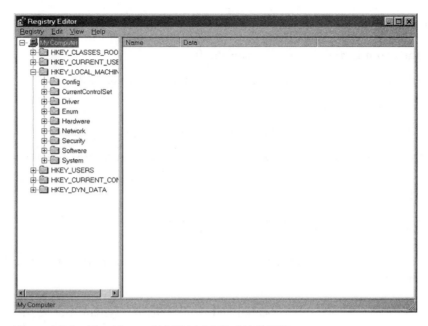

Figure L8.2   The Hives of HKEY_LOCAL_MACHINE

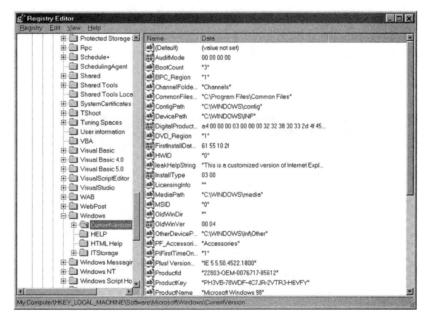

Figure L8.3   The Assorted Values of CurrentVersion

6.   To see the type of data that is stored in a particular entry, double-click FirstInstallDateTime. Hmmm. They seem to have encrypted this. Wonder why they might have done that?

7.   So far, you haven't changed anything in the Registry (or at least I hope you haven't). Before you do anything that drastic, you want to have a backup of the Registry. To do that, highlight My Computer at the very top of the navigation area. Then click Registry → Export Registry File. The default location to save the file is in the My Documents folder. Accept the default folder and name your backup **regbackup.reg**. If you screw anything up beyond recognition, you can get your system back.

EXERCISE 1 REVIEW

1. What are the two primary files that contain the Registry?

2. What additional file is created if you choose to enable system policies?

3. List the six primary keys of the Registry.

4. Where would settings specific to hardware installed on the system be stored?

5. Where would you look for the settings that dictate how a particular file behaves if double-clicked?

EXERCISE 2: EDITING THE REGISTRY

If you haven't made your Registry backup yet, *do not continue.* Go back to the last step in Exercise 1 and complete it before going on.

1. You want to make a change that is benign and won't hurt the system. But at the same time, you want it to be something for which the results of your edit will be noticed. Right-click the Start button and click Explore. This will open Windows Explorer.

2. Click View → Folder Options. This will bring up the window shown in **Figure L8.4**. Note the line that reads "Remember each folder's view settings." You're about to change that.

3. In the Registry editor, open the HKEY_LOCAL_MACHINE → Software → Microsoft → CurrentVersion → explorer → advanced → folder. Highlight ClassicViewState in the

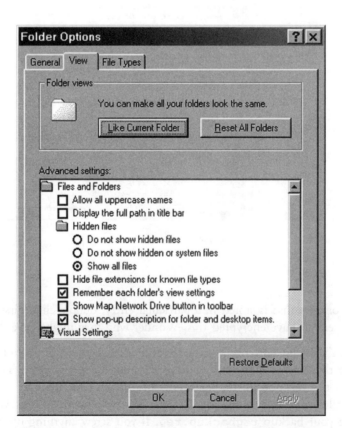

Figure L8.4   Folder Viewing Options in Windows Explorer

navigation area. On the topic area, several entries will appear. Double-click Text. In the Edit String field that appears, type the words **These are Equal Opportunity Folders** and click OK.

4. Close and restart Windows Explorer. It is not necessary to restart the machine. Now when your view folder opens how does that entry read?

EXERCISE 2 REVIEW

1. What is the first thing you should do before messing with the Windows Registry?

2. What are two results that can occur from inadvertently deleting an entry in the Registry that wasn't supposed to be deleted?

EXERCISE 3: THE SCANREG UTILITY

Of course, now you want to put things back the way they were. Obviously, the easy way is to browse to that Registry entry and type the words the way they used to be. But what if you can't remember how it used to read? Or, what if you can't remember just what entry you edited? Or worse yet, what if Windows blue-screens just as it's entering the graphical mode? There's got to be a way of rolling back the system to a previous setting. In fact, there are two ways. The first is, of course, to open the Registry editor and click <u>R</u>egistry → Import Registry File and then select the backup of the Registry that you made. But that's too easy and doesn't give me the chance to show you the ScanReg utility. Windows 2000 and XP users, unfortunately, don't have this utility at their disposal.

1. Restart your machine and as POST is completing, just before Windows starts to boot, press the F8 key. This will bring up a boot menu. One of the options is to start Windows to a Command Prompt. Select this option and continue booting. This will bring you to a C:\ prompt.

2. Type **scanreg** at the command prompt. The ScanReg utility will start, and the first thing it will do is to scan the Registry for errors. You can't stop this process despite the fact that, in all the years I've used this utility, no matter how corrupted the Registry was, ScanReg always reported no errors.

3. Next, you are prompted to either back up your Registry or to view existing backups. Obviously, if you're having a problem with the current version of the Windows Registry, the last thing you want to do is make a backup of a corrupted Registry. Tab over to View Backups. You will see four versions of Registry backups listed by the date they were created. Pick the backup previous to the one with today's date and select Restore. After a few moments, it will prompt you to restart your machine. Windows will boot to the older backup without the changes you made.

EXERCISE 3 REVIEW

1. What is the first thing that ScanReg prompts you to do?

2. What is a key difference between running ScanReg from the GUI and running it from the command prompt?

LAB REVIEW

1. Which Registry key is not stored in a file, but rather is created on the fly as the system boots?

2. What are two different ways you can use to back up the full Windows Registry?

3. You have just installed a new video card. During the driver installation you asked to see a list of all devices. You're so used to another computer that you use every day that you mistakenly selected a completely different make and model of video card. Now every time you boot your machine, as soon as the Windows graphical interface starts to load, the screen goes blank. How might you fix this? (There are actually a couple of ways. If you come up with a way I didn't discuss in this lab, you should lobby the instructor for extra credit.)

4. Which primary key holds the user settings for every user with an account on the system?

5. How many backups of the Registry does Windows 98 maintain?

Lab Summary

One of the things I hope you learned from this lab is that every aspect of how Windows performs is a function of some entry in the Registry. In Lab Seven you learned how to make safe changes to the Registry in Control Panel. Here, I showed you the basics of how the Registry Editor can be used to make changes beyond the scope of Control Panel. At the risk of sounding like a skipping CD, before you mess around with the Registry, *back it up!!!*

A Tour of the BIOS

Because the system BIOS is responsible for starting the computer, making sure that the system devices all work properly, and speaking to the CPU on its terms, it's probably a good idea if the average technician knows how to configure it properly. As I discussed in Chapter Six of the textbook, Motherboards, BIOS, and the Chipset, three executable programs reside on the BIOS. These programs are Power On, Self Test (POST), Setup, and Bootstrap Loader. What I'll be doing in this lab is pointing out some of the more common settings of Setup. For more information on POST and Bootstrap Loader, refer to Chapter Six of the textbook. For this lab, all you need is the student system equipped with keyboard and monitor. You won't even need the mouse. CompTIA exam objectives covered in this lab include the following:

1.1 Identify the names, purpose, and characteristics of system modules. Recognize these modules by sight or definition.

2.1 Recognize common problems associated with each module and their symptoms, and identify steps to isolate and troubleshoot the problems.

4.4 Identify the purpose of CMOS (Complementary Metal-Oxide Semiconductor) memory, what it contains, and how and when to change its basic parameters.

One thing I should point out before you proceed is that I will be using an Award BIOS in the lab. Machines with an AMI BIOS should be very similar, but there will be some differences as to what settings might and might not appear. In fact, you might have an Award BIOS and get a somewhat different screen. However, you will see enough of the settings I describe, and their descriptions will be similar enough, to make the exercise worthwhile. If the machines you use are equipped with Phoenix BIOS, another brand not mentioned, or a BIOS generated by a major computer manufacturer, you will most likely find following this lab step by step inconvenient and confusing. Just go through the various menu options on your system, and try to associate the entries you find with the ones in this lab whenever possible.

EXERCISE 1: A TOUR OF THE BIOS

1. First make sure that the monitor is on and running before starting the machine. Monitors frequently take longer to get up and running than the amount of time you're given to press the key sequence needed to enter the Setup program.

2. Turn on the system and watch carefully. Somewhere on the screen, usually in either the upper right-hand corner or centered near the bottom of the screen, will be a message telling you what

key to press to enter the Setup program. If you are using proprietary machines with a splash screen that appears on startup, you usually can exit the splash screen by pressing the <Esc> key. If your machine gives no indication of what keys to press, try some of the following:

a. <Delete>: Award, AMI, and some proprietary

b. <F2>: Dell and some Phoenix

c. <F10>: Compaq and some Phoenix

d. <Ctrl> + <Alt> + <Delete>: Some proprietary

e. <Esc>: Some proprietary

3. Those of you running Award or AMI will be rewarded with a screen that looks very much like the one in **Figure L9.1**. Depending on the brand of motherboard and what version your BIOS is, the categories may vary.

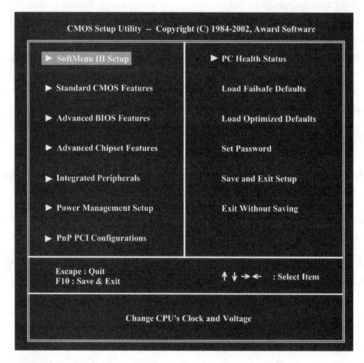

Figure L9.1 The Opening Screen for an Award BIOS

4. For the remainder of this exercise, I'll go through the categories one by one and explain the meaning of each setting and the effect it has on the machine, including any ill effects of an incorrect setting. Because so many different people will be seeing so many different screens, I won't be using screenshots for the remainder of this lab. It would just add confusion for those not using Award BIOS.

■ *Soft Menu III Setup:* This menu option will appear only for motherboards that support BIOS configuration of CPU parameters. The four main user-configurable parameters are as follows:

a. CPU Operating Speed: Some subparameters listed below this option are grayed out. This is because Intel has locked certain parameters, including CPU multiplier. However, the CPU can be set to run beyond or below its rated speed. Changing CPU operating speed will affect some of these parameters. A change that defines a multiplier not supported by the CPU will prevent the system from booting, and in most cases, automatically start Setup during the next attempt so that you can fix the problem.

b. DRAM Clock: The options on this particular version are Host Clock and HCLK-PCICLK. The Host Clock is the speed of the front-side bus (FSB). The PCI clock is 33Mhz. If the FSB is set at 133Mhz in the CPU Operating Speed, then selecting Host Clock means that your DRAM will operate at that same speed. Setting it at HCLK-PCICLK means that RAM will operate at the host clock speed *minus* the PCI clock speed, or 100Mhz in this case. This enables you to use memory designed for slower systems on this computer.

c. CPU Power Supply: The options are CPU Default or User Defined. CPU Default auto-detects two settings, Core Voltage and VCC3 voltage. Core Voltage is the voltage that powers the CPU, and VCC3 voltage is the voltage that exists between the CPU and I/O. Typically Core can range from around 1.3V to 2.8V. This particular board supports 1.3V to 1.85V. Unless you are attempting to overclock your system, these settings should remain at the default settings. However, if you attempt to run your CPU beyond rated speed, it might be necessary to bump these settings up a notch. Be forewarned that doing so can make your system unstable.

d. CPU Hardwired IRQ: By default PCI devices have four IRQ channels. The options here are 4 and 1. The default is 4 and should remain that way unless someone at the manufacturer's technical support line instructs you to change it. Setting it to 1 will cause the system to check only one channel during POST. If that channel's IRQ is already taken, it will record a conflict, and the device won't be recognized. Setting it to 4 forces POST to scan all four channels.

■ *Standard CMOS Features:* Here, features common to all computer systems are recorded. With a couple of minor possible exceptions, nothing in here will prevent your system from booting if a setting is incorrect.

a. Date: Sets the date.

b. Time: Sets the time.

c. IDE Primary Master: Auto, User Defined, None. Auto lets the system query the hard disk for its parameters and adjusts the settings accordingly. User Defined enables the user to input the number of cylinders, heads, sectors/track, and a predefined landing zone. None shuts off that particular channel, and no device attached to it will be recognized.

d. IDE Primary Slave: See previous.

e. IDE Secondary Master: See previous.

f. IDE Secondary Slave: See previous.

g. Drive A: Your primary floppy drive. Options may include 360KB 5.25-inch, 1.2MB 5.25-inch, 720KB 3.5-inch, 1.44MB 3.5-inch, and possibly 2.88MB 3.5-inch. An option for None is also given. On some versions of BIOS, None is the default option, and the system will revert to None if power is cut off from the CMOS chip for too long. Setting this parameter to None will cause the system to ignore the floppy on bootup, making it inaccessible to the user. Setting the parameter incorrectly will cause errors in POST, and if you choose to continue, the floppy will not be readable when the system is booted.

h. Drive B: Your secondary floppy drive (if installed). See previous.

i.   Floppy 3 Mode Support: This is a Japanese standard that stores 1.2MB on a standard 3.5-inch high-density floppy diskette. If you have a floppy drive that supports this mode and you want to enable it, the options are None, Drive A, Drive B, and Both.

j.   Video: The options are EGA/VGA, CGA40, CGA80, and Mono. I know of no monitor manufacturers still making CGA or the old-style Hercules mono-mode monitors. Unless you managed to stumble across one of these archeological finds and have some masochistic desire to actually *USE* it, the default setting of EGA/VGA should not be changed.

k.   Halt On: Defines what POST errors can be ignored. Typical settings are None, All But Keyboard, All But Diskette, and All But Disk/Key. Keyboards occasionally exhibit "phantom errors." Even if the keyboard is not found, it may be preferable to let the system boot. Many systems allow a PS2 or a USB keyboard to be inserted on the fly after POST is completed. Some people rarely, if ever, use their floppy disk drive. Having their entire machine become unusable because that drive has failed would be unacceptable.

■   Now I'll move on to the Advanced BIOS Features. Most of the settings are safe enough and cause no more harm than seriously reduced system performance if incorrect. A couple of features that I'll point out can possibly prevent your system from booting. If that happens, restart the machine, enter Setup, and restore the system to Factory Defaults. I'll be covering that a little later in this lab.

a.   Virus Warning: Enabled/Disabled. Prevents the system from making changes to either BIOS code (if the BIOS is flashable) or the hard disk's master boot record. This may need to be disabled when reinstalling an operating system.

b.   Cache Level 1: Enabled/Disabled. Enables read/write operations to L1 cache built into the CPU. Disable *only* if you know you have a CPU such as the original Celeron, which has no L1 cache onboard.

c.   Cache Level 2: Enables read/write operations to L2 cache built into the CPU. On older motherboards, the L2 might be located on the system board. Disable *only* if you know you have a CPU such as the original Celeron, which has no L2 cache onboard.

d.   CPU Level 2 Cache ECC Check: Enabled/Disabled. Turns on/off with the use of the Error Correction Code.

e.   Quick Power On Self Test: Enabled/Disabled, Full Test. Tests only selected system components on cold boot.

f.   HDD Sequence, IDE/SCSI, Where to look for MBR Boot Sequence: Options vary. Determines the order of devices in which POST looks for the MBR.

g.   Boot Up Floppy Seek: Enabled/Disabled. Tests floppy drive to see whether it has forty or eighty tracks.

h.   Floppy Disk Access Control: R/W/Read Only. Security access for floppy drive.

i.   HDD S.M.A.R.T capability: Enabled/Disabled. If your hard disk supports a Self Monitoring And Reporting Technology (and everything made for the past several years does) option, then this should be enabled. To take advantage of it, you'll need to install a utility on your system as well.

j.   PS2 Mouse Function Control: Enabled/Auto. Looks to PS2 for the mouse.

k.   OS2 Onboard Memory: Enabled/Disabled. Use memory mapping functions as defined by IBM's OS2 operating system. Unless you are actually running a copy of that now-defunct OS, this should be disabled.

l. Video ROM/BIOS Shadowing: Enabled/Disabled. Allows copying of BIOS routines from your video card to upper memory for enhanced performance. Enabled is best unless you are told otherwise by your manufacturer.

m. C8000-DFFFF Shadowing (multiple entries): Enabled/Disabled. Allows copying of Supplemental BIOS routines of specific devices to specific addresses. Enable *only* if the installation instructions for a specific device instructs you to do so. In that case, the manufacturer will also let you know which address range (or ranges) to use.

n. Boot Up NumLock Status: On/Off. Determines whether number lock on keyboard is on or off after system boots.

o. Typomatic Rate Setting: Disabled/Enabled. Disabled turns off Typomatic Rate and Typomatic Delay.

p. Typomatic Rate: Options vary. Sets speed at which characters repeat when a key on the keyboard is held down.

q. Typomatic Delay: Options vary. Sets time that elapses before keys begin to repeat when a key on the keyboard is pressed.

r. Security Option: System/Setup. Determines to what any security settings you configure will apply. System dictates that a password will be needed to boot the system. Setup enables the user to boot the system without a password but requires a password to run the CMOS Setup Utility (the one you're currently examining). Security Settings are found elsewhere in the BIOS.

■ Now I'll enter the possibly intimidating and potentially dangerous arena of the Chipset Features Setup. This is where an incorrect setting can render your system useless until corrected. As with the Advanced BIOS Features, you simply reboot, reset to Factory Defaults, and start all over.

a. EDO Autoconfiguration: Enabled/Disabled. Enables chipset to control timing functions for Extended Data Out (EDO) memory. This and the other EDO-specific functions will not appear on a system that does not support EDO. (That shouldn't be much of a surprise.)

b. EDO Read Burst Timing: Varies. Sets number of clock cycles for Burst Mode read operations. Setting this or any of the following memory timing parameters too fast can prevent the system from booting. Setting them too slow will theoretically hinder performance.

c. EDO Write Burst Timing: Varies. Sets number of clock cycles for Burst Mode write operations.

d. EDO RAS Precharge: 3T/4T. Sets the number of clock cycles for RAS Precharge.

e. EDO RAS/CAS Delay: 2T/3T. Sets the number of clock cycles for RAS/CAS Delay.

f. SDRAM Configuration: Varies. Sets clock speed of SDRAM.

g. SDRAM RAS Precharge: Auto/3T/4T. Sets the number of clock cycles for RAS Precharge.

h. SDRAM RAS/CAS Delay: Auto, 3T/2T. Sets the number of clock cycles for RAS/CAS Delay.

i. Graphics Aperture Size: Varies. Setting this option too low can result in degraded video performance; however, setting it too high takes memory away from applications running on the system should another application take full advantage of the setting.

j.   PCI 2.1 Support: Enabled/Disabled. Disabled setting drops system back to PCI Version 1.0.

k.   Memory Hole at 15M–16M: Enabled/Disabled. ISA Devices can read up to only 16MB of RAM because of their 24-bit address space. Some very old ISA devices have compatibility issues with extended memory beyond 16MB, and the area between 15MB and 16MB needs to be reserved as a buffer *window* for bring data down from that area. Unless you are using one of these older cards, this option should be disabled.

l.   Onboard FDC Controller: Enabled/Disabled. Enables you to disable the floppy disk drive.

m.   Onboard Floppy Swap A/B: Enabled/Disabled. Has no effect on which floppy drive is bootable, but after POST is completed, it switches Drives A and B so that the primary drive is listed as B and vice versa.

n.   Onboard Parallel Port: Various settings/Disabled. Enables disabling or reconfiguring of the parallel port to a different LPT port.

o.   Parallel Port Mode: Normal, ECC, ECC/ECP. Sets up parallel communications.

p.   ECP DMA Select: Varies. Sets DMA channel used by ECP Parallel mode. Grayed out if a parallel mode other than ECP is configured.

q.   UART2 Use Infrared: Enabled/Disabled. Sets infrared port to UART2.

r.   Onboard PCI/IDE Enable: Both, Primary, Secondary, Disabled. Enables/disables IDE ports.

s.   IDE DMA Mode: Auto/Disable. Disables, autoselects Direct Memory Access mode for IDE devices.

t.   IDE 0/1—Master/Slave: Various. Sets PIO mode and DMA channel for specific device.

u.   CPU/PCI Write Buffer: Enable/Disable. When enabled sets up a memory cache for data being moved from the PCI bus to the CPU. Unless otherwise specified by the manufacturer, this should be enabled.

v.   PCI Dynamic Bursting: Enable/Disable. When enabled, combines multiple 8- or 16-bit write executions into 32-bit operations. Some older PCI NICs are not compatible with this option, but otherwise, this setting should be enabled for best performance. If your NIC is incompatible, it won't work, so this is one area to look at when troubleshooting an older PCI NIC you just installed.

■ Integrated Peripherals: Here is where many of the settings for devices such as the IDE ports, floppy ports, and serial ports are set. Systems with embedded peripherals such as onboard sound and video will very likely see a number of other settings here as well.

a.   Onboard IDE-1 Controller: Enabled/Disabled. Enables the user to turn the controller off or on as required.

   i.   Master Drive PIO Mode: Modes 0-4. Sets Programmed Input/Output mode for the device on this channel.

   ii.   Slave Drive PIO Mode: Same as previous.

   iii.   Master Drive DMA Mode: Modes 1-5. Sets Ultra DMA Mode for the device on this channel.

   iv.   Slave Drive DMA Mode: Modes 1-5. Same as previous.

b.   Onboard IDE-2 Controller. All of the parameters and subsequent parameters for PIO and DMA modes are the same as previous.

c.   IDE Prefetch Mode: Enabled/Disabled. On drives that support this function, it allows the controller to initiate the read process for the next sectors of data that are likely to be needed. If a hard disk is constantly generating drive read errors after the boot process is completed, try disabling this setting.

   d. Init Display First: PCI/AGP. If two graphics adapters are installed in the computer, this setting determines which is the primary and which is the secondary adapter.

   e. USB Controller: Enabled/Disabled. Turns the Universal Serial Bus controller on/off.

      i. USB Keyboard Support: OS/BIOS. Determines whether a USB keyboard will be managed by the BIOS or the operating system.

   f. HDD Block Mode: Options vary. Dictates the number of sectors a hard drive can transfer during each interrupt request. With modern drives, the maximum setting is reported by the drive during POST, and the value should not be set higher than the default. It can prevent the system from booting. A lower setting can resolve buffer overflows.

   g. Onboard FDD Controller: Enabled/Disabled. Turns on/off the floppy disk drive controller as needed.

   h. Onboard Serial Port 1: Various COM port settings/Disabled. Allows disabling or reconfiguring Serial Port 1 to a different COM port.

   i. Onboard Serial Port 2: Various settings/Disabled. Allows disabling or reconfiguring Serial Port 2 to a different COM port.

■ Power Management Setup: These days conserving energy is of concern to everyone, or at least it should be. A company that runs its computers 24 hours a day, seven days a week, even when users are available only eight hours a day, five days a weeks, wastes a tremendous amount of energy. Configuring a computer to automatically shut down the entire system, or individual components when not in use, can have a noticeable impact on the bottom line.

   a. Power Management: User Defined/Max/Min. User Defined enables users to input shutdown times for each individual component as they see fit. Max provides settings for maximum energy savings and shuts the system down after a very short period. The actual number of minutes can vary between brands of BIOS but is generally between one and three minutes. Min generally sets the shutdown for all components at thirty minutes.

   b. PM Control by APM: Yes/No. Stands for Power Management by Advanced Power Management. APM is a Microsoft/Intel software implementation of power management and turns control over to the operating system.

   c. Video Suspend Option: Always On/Suspend 'off/All Modes' off. Different modes are Suspend, Standby, and Hibernation (on notebooks). Determines which mode the system will go into when the configured time of inactivity has elapsed.

   d. Video Off Method: Blank Screen/V/H Screen +Blank/DPMS Support. Blank Screen only blanks the screen. V/H Screen +Blank will turn off the V-Synch and H-Synch signals traveling between the video card and the monitor as well as blanking the monitor. Display Power Management System is a software controlled method, which, if supported by the monitor, allows the operating system to establish values.

   e. Modem Uses IRQ: N/A or a listing of IRQ Settings. On a system that supports Wake on Modem, the IRQ of the modem needs to be configured here in order for the service to work.

   f. Soft-off By Pwr Button: Instant-off/Delay/Standby. Determines the effect of pressing the power button. Instant-off immediately turns the system off. Delay requires the user to hold the button in for at least four to six seconds before the system will shut down. Standby enables the user to press the button briefly to put the system into standby or to hold it in four to six seconds to shut the system off.

g. State After Power Failure: Off/On. If set to off, when power is restored, the system will remain off. When set to on, it will reboot.

h. Wakeup Events: Various settings.

    i. VGA: Off/On. If set to On, a signal from the VGA card can wake the system.

    ii. LPT and COM: None/LPT/COM/LPT-COM. None prevents a signal from either LPT or COM from waking the system. The other three enable the user to set either LPT, COM, or both LPT and COM as signals that will wake it.

    iii. HDD and FDD: On/Off. Activity from either of these drives will wake the system when set to On.

    iv. PCI Master: On/Off. PCI Master refers to any device on the PCI bus that can initiate a data/exchange sequence. Data coming in from one of these devices will wake the system when set to On.

    v. Power on by PCI Card: On/Off. Allows incoming data from a PCI card to activate the system.

    vi. Wake on LAN/RING: On/Off. When enabled, allows either the modem or the NIC to activate the system.

    vii. RTC Alarm Resume: On/Off. When set to On, allows the Real Time Clock to wake the system. This will cause the next two values to become configurable.

        1. Date (of month): 1-31. The system will automatically activate each month on the date set.

        2. Resume Time: (hh:mm:ss). The exact time at which the system will resume. If a day is set in the Date field, the system will resume at that time on that date, each month. If no date is set, the system will wake up at that time every day.

    viii. Primary INTR: On/Off. Allows enabled IRQ channels to activate the system.

    ix. IRQs Activity Monitoring: Press Enter. This is where you enable or disable IRQs for the previous setting.

■ The next section is entitled PnP/PCI Configurations. This can be a handy part of the BIOS to understand when you have a persnickety adapter card that won't behave. Nothing in here will prevent your system from being usable if incorrectly configured, but an incorrect setting might make a particular device disappear from the system.

a. Reset Configuration Data: Disabled/Enabled. This option forces the Extended Systems Configuration Data (ESCD) to be rebuilt from scratch. During the POST that follows the enabling of this option, all Plug 'n Play devices will be forced to relinquish their assigned resources, and new allocations will be assigned based on the now-current configuration. This option should be set to Enabled any time changes in the remaining fields on this page are made. With most (if not all) brands of BIOS, this parameter will reset itself to Disabled after the reconfiguration has been completed.

b. Resources Controlled By: Auto(ESCD)/User. When set to Auto(ESCD), Plug 'n Play manages all resources. When set to User, the following setting will be enabled.

c. IRQ Resources: Press Enter. All available IRQs, and with certain brands of BIOS, the DMA channels as well, will appear in a list. The options for each entry are PCI Device or Reserved. Some version use ISA in place of Reserved. If set to PCI Device, that resource is kept available to Plug 'n Play. If not, it is removed from the available resource list, and a non-Plug 'n Play device won't have to compete for it (and usually lose).

d. PCI/VGA Palette Snoop: Disabled/Enabled. Allows video adapters to directly access RAM looking for video information. Unless you're using a PCI video card, this should be disabled.

e. Assign IRQ for VGA: Enabled/Disabled. Most modern VGA cards require an IRQ and, therefore, this needs to be enabled.

f. Assign IRQ for USB: Enabled/Disabled. If you do not use USB devices, there is no point in tying up the IRQ assigned to the USB hub on your machine. However, if this is disabled and you attempt to install a USB device, it won't be recognized.

g. PCI Latency Timer CLK: 0-255. Sets the number of FSB clock cycles that a PCI Master device can maintain control of in the PCI bus. Setting this number too high takes clock cycles away from the CPU. Setting it too low can impede PCI I/O operations.

h. PIRQ_0 User IRQ: None/0-15. Configurations for Programmable Interrupts. Unless you *really* know what you're doing, this is not a setting you want to play with.

i. The next three settings are the same as the preceding for PIRQ_1 through PIRQ_3.

■ Load Failsafe Defaults: Y/N. Okay, you messed around with some of those advanced chipset parameters I warned you about, and now your system won't boot. Remember that earlier I mentioned I would show you how to revert to factory defaults. In this version of BIOS there are actually two collections of default settings. Failsafe Defaults is the setting to use when all else fails. It disables all but the minimum configuration settings required to get the system to boot.

■ Load Optimized Defaults: Y/N. As with the previous setting, this loads a collection of default settings, but instead of the bare minimum settings required to boot, the Optimized Defaults are the settings designed for the greatest efficiency of the motherboard. This is the default setting you should always try first if playing around in the BIOS has rendered your system unbootable.

■ Set Password: Opens a blank field into which you can type a password. The security setting I discussed earlier determines whether the password applies to booting the system or to entering the CMOS Setup program.

■ Save and Exit Setup: Y/N. Y saves all the configuration changes you've made in your session, exits setup, and reboots the machine.

■ Discard Changes and Exit Setup: Y/N. Throws out all the changes you made and reboots the machine back to the way it was before you started playing around.

■ Select Discard Changes and Exit Setup, press <Enter>, and let your system reboot. You're finished with this exercise. (Whew!)

EXERCISE 1 REVIEW

1. Using what you learned from the preceding exercises and from your textbook, explain why changing a few settings in the Advanced BIOS Settings is unlikely to harm your computer, while changes in Advanced Chipset Settings may prevent your system from booting.

2. You've just made several changes in your BIOS settings, and now the system won't boot. What are your options?

3. When trying to fine-tune a system, why do you think that seasoned technicians always follow the rule that states, "Make only one change at a time."

4. Which menu would you most likely open if you wanted to check how your system was configured for RAS/CAS Delay?

5.   A user in your company has left without notice and had password encoded the system she was using before her untimely departure. How are you going to get back into that system?

EXERCISE 2: BACKING UP THE CMOS

Well, you can see that in a finely tuned system, quite a number of changes can be made in Setup. It's safe to assume that six months from now, after you've changed your battery and discovered that your CMOS has been wiped clean, you're going to remember every single change you made and precisely what those changes were, right? If you can do that, can I talk to you about working with me on my next book?

What you want is a safe way to back up those settings after you've made them. Notice that nowhere in any of the preceding sections were you offered an opportunity to do a CMOS backup. With a couple of notable exceptions (such as Compaq), CMOS backup is not an option. You have two choices when it comes to backing up your CMOS. You can either record every change you make on a sheet of paper and hope you don't lose that paper between now and the time you actually need it, or you can use one of the several CMOS backup utilities available. These utilities range from free to relatively expensive. Because I'm notoriously cheap, I'm going to show you one of the freebies that I stumbled across that hasn't failed me yet. It's called CMOSBAK.EXE and is available at www.mwgraves.com.

When you first download this utility, it's going to be in zipped form, so you'll need an unzipping utility to expand it. I use Ultimate Zip, another freeware utility available for download on the previously mentioned site. Assuming that you haven't downloaded either, getting a useful rendition of CMOSBAK will involve the following steps.

EXERCISE 2A: ADDING CMOSBAK TO THE TECHNICIAN'S BOOT DISK

1.   On a computer with Internet access, log on to www.mwgraves.com, click the Downloads link, and download both the Ultimate Zip and the CMOSBAK utilities. Ultimate Zip is a self-extracting executable and will unzip and install itself when you run the file.

2.   When Ultimate Zip is installed, create a directory on your hard drive called CMOSBAK, and unzip CMOSBAK.ZIP to that directory.

3.   Copy only the file CMOSBAK.EXE onto the Technician's Boot Disk that you made in Lab. You now have an improved model of the Technician's Boot Disk. And don't get too excited. Before you've finished with this course, it will be even better.

EXERCISE 2A REVIEW

1.   Use your imagination. Why do you think CMOSBAK is designed to run from the command prompt?

EXERCISE 2B: RUNNING CMOSBAK

Now that you've got CMOSBAK added to your disk, you're ready to back up the CMOS on your system.

1.   Insert the Technician's Boot Disk into the floppy drive and start your machine. Once again, it will boot to the A: prompt.

2.   From the A: prompt, type CMOSBAK. You'll get a screen like the one shown in **Figure L9.2**.

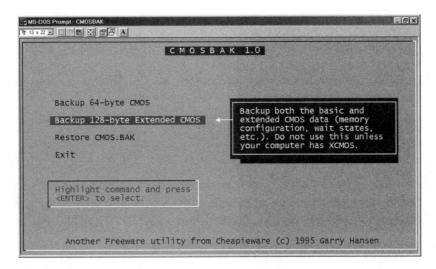

Figure L9.2 CMOS.BAK is a utility that allows the user to back up CMOS settings to a floppy disk.

3. Unless you have a very old machine, select the option to Backup 128-byte Extended CMOS. Even if you're wrong, you won't do any harm. You'll have to use the Down Arrow <↓> on your keyboard to select this option. The mouse won't be working at this point.

4. Press the <Enter> key, and in a second or less a message will tell you that your CMOS has been successful backed up.

EXERCISE 2B REVIEW

1. CMOSBAK.EXE always creates files with the same name. How can you go about using it to back up a number of computers on a single floppy?

EXERCISE 2C: RESTORING THE CMOS

That was easy, wasn't it? Now let's see how to put our systems back to the way you had it. You actually want to see that something really happened, don't you? So let's make some changes in the CMOS before you use the Restore function of CMOSBAK.

1. Remove the Technician's Boot Disk from the floppy drive and restart your machine. This time you want to press the appropriate key sequence to enter the Setup program.

2. Navigate to the Boot Sequence in your CMOS and rearrange the order of the sequence. Record your changes on a sheet of paper.

3. Now find the settings for the serial and parallel ports and (if your BIOS supports making this change) set the COM and LPT settings to different settings. Record your changes.

4. Save your changes and exit. As the machine reboots, re-enter the Setup program and verify that your changes stuck.

5. Insert the Technician's Boot Disk into Drive A and restart the machine.

6. From the A: prompt, type **CMOSBAK**.

7. Select the Restore CMOS.BAK option. In a second or two, it will inform you whether the restore was successful or not. Remove the disk and reboot the machine, once again entering the Setup program.

8. Review the changes you made and verify that they are back to their original settings.

EXERCISE 2 REVIEW

1. What is the purpose of backing up the CMOS after achieving the optimum configuration?

2. One of the shortcomings of CMOSBAK is that a single disk can back up the CMOS settings for only a single computer. Based on what you saw during the Restore exercise, explain why this is so.

3. Why doesn't the mouse work while running the CMOSBAK utility?

LAB SUMMARY

If there is any one of the labs in this manual you might want to do several different times, using several different machines, it is this one. Being able to navigate the Setup program of different versions of BIOS is one of the most critical things a technician can learn. Unfortunately, although there is a great deal of similarity between different brands of BIOS, there are even more differences. It's a good idea to get to know as many different brands as you can.

RUNNING DIAGNOSTICS

I've pointed out time and time again to students and readers that troubleshooting is both a science and an art. When it comes to the art, you either have it or you don't. Many people are able to develop it over time with practice. The science, however, is something I can teach you. In the first half of this lab, I'm going to lead you through some of the diagnostics procedures using some of the freeware diagnostics utilities that are available. The second section of this lab is optional, depending on whether or not your class is able to provide a POST Card for the procedures. To finish both of these labs, you will need the following:

- The student computer for each student, booted to Windows

- A blank, formatted 3.5-inch 1.44MB floppy disk

- TuffTEST Lite (latest version; this lab was written with v1.53)

- For the optional lab, the classroom will need at least one POST Card hardware diagnostics tool. Any brand will do. (This lab was written using the Ultra-X PCI POST Card.)

- A pencil and paper

The CompTIA exam objective covered in these labs is the all-important

2.1 Recognize common problems associated with each module and their symptoms, and identify steps to isolate and troubleshoot the problems.

Get your stuff together, and let's get started.

EXERCISE 1: SOFTWARE DIAGNOSTICS

For this exercise, I have chosen TuffTEST Lite for you to examine. Although it may not be the most potent piece of diagnostics software available, it has the advantage of being free and does perform the diagnostics I want to demonstrate in this lab. Should you choose to purchase the Professional version later, some of the glaring limitations of the free version are eliminated, and you will have a potent contender.

1. Start by downloading and installing TuffTEST Lite onto your PC. When prompted, insert your blank, formatted disk into the drive and make the bootable floppy as prompted by the installation routine. This application cannot be added to your Technician's Boot Disk (TBD), because

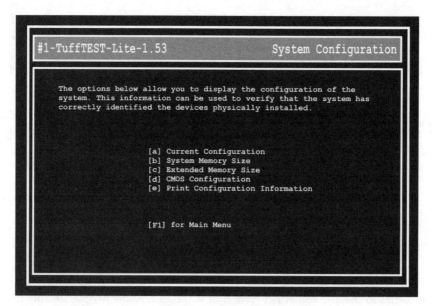

Figure L10.1   The View Configuration Page of TuffTEST Lite

the tests must be run independently of any OS. Your TBD loads a basic Windows kernel. As such, devices dependent on Windows drivers to function might not be tested.

2. After the bootable floppy has been completed, restart your machine with the floppy still inserted in the drive. There will be two introductory screens, one introducing the product and another promoting the Single-user and Professional versions of their more sophisticated products. Next comes the main menu. You have precisely 10 seconds to decide what you want to do on this screen, or it will automatically assume that you want to run each and every test available. You don't, so press F2. A screen similar to the one shown in **Figure L10.1** should appear.

3. Note that the options are to view the following:

    a. Current Configuration

    b. System Memory Size

    c. Extended Memory Size

    d. CMOS Configuration

    e. Print Configuration information

    f. Press F1 to return to the main menu

4. After viewing each of the preceding options, press F1 to return to the main menu. Now you've got more time to explore this menu (**Figure L10.2**).

5. The first option, F2, brings you back to the configuration menu that you just examined. F3 enables you to pick and choose the different tests that you want to perform. These options include the following:

    a. System Board Tests

        i.   Microprocessor

        ii.   Math Coprocessor

        iii. DMA Controller

        iv.   Real Time Clock

    b. Video Alignment Aids

        i.   Cross-hatch Pat(tern)

        ii.   Dot Pattern

        iii. Vertical Bars

        iv.   Horizontal Bars

        v.   Text Color Chart

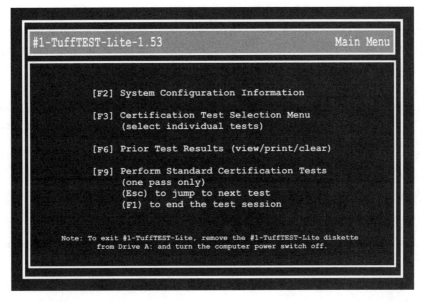

```
┌─────────────────────────────────────────────────────────┐
│ #1-TuffTEST-Lite-1.53                          Main Menu │
├─────────────────────────────────────────────────────────┤
│                                                          │
│       [F2] System Configuration Information              │
│                                                          │
│       [F3] Certification Test Selection Menu             │
│            (select individual tests)                     │
│                                                          │
│       [F6] Prior Test Results (view/print/clear)         │
│                                                          │
│       [F9] Perform Standard Certification Tests          │
│            (one pass only)                               │
│            (Esc) to jump to next test                    │
│            (F1) to end the test session                  │
│                                                          │
│                                                          │
│     Note: To exit #1-TuffTEST-Lite, remove the #1-TuffTEST-Lite diskette │
│           from Drive A: and turn the computer power switch off.           │
│                                                          │
└─────────────────────────────────────────────────────────┘
```

Figure L10.2   TuffTEST Main Menu

c. Video Adapter Tests

   i.   Verify Screen Mem(ory)

   ii. Character Set

   iii. Video Attributes

   iv. Fill Display

   v.   Cursor Addressing

   vi. Video Memory

   vii. High-Res Color

   viii. Text Resolution

d. Parallel Port Tests

   i.   LPT1 (I/O Address)

   ii. LPT2 (I/O Address)

   iii. LPT3 (I/O Address)

e. Serial Port Tests

   i.   COM1 (I/O Address)

   ii. COM2 (I/O Address)

   iii. COM3 (I/O Address)

   iv. COM4 (I/O Address)

f. Diskette Drive (A:) Tests

   i.   Reset Drive

   ii. Rotational Timing

   iii. Seek/Read

   iv. Read Only

g. Diskette Drive (B:) Tests

   i.   Reset Drive

   ii. Rotational Timing

   iii. Seek/Read

   iv. Read Only

h. Fixed Disk (0) Tests

   i.   Controller

   ii. Seek (Hysteresis)

   iii. Surface Test

i. Fixed Disk (0) Tests

   i.   Controller

   ii. Seek (Hysteresis)

   iii. Surface Test

j. Main Memory Tests

   i.   All Zeros

   ii. All Ones

   iii. Checkerboard

   iv. Address

   v.   Walking Ones

k. Extended Memory Tests

   i.   All Zeros

   ii. All Ones

   iii. Checkerboard

   iv. Address

   v.   Walking Ones

6. You can toggle individual tests on or off by typing the number specific to that test, as viewed on the screen. Entire groupings of tests are toggled on and off by typing the letter for that test. To run all tests from this menu, press F8. As you do so, read the following text explaining these tests.

- *CPU:* In this particular program, the manufacturer does not specify what tests are being performed. I was, however, able to determine that, at the minimum, a test for the accuracy of speed is performed.

- *Math Coprocessor:* Tests the MathCo section for floating-point errors.

- *DMA Controller:* Queries and checks response to each DMA channel in the system.

- *Real Time Clock:* Queries and checks response to RCT and alarm functions.

- *Cross-hatch Pattern:* Tests the monitor's capability to display images created by multiple crossing lines.

- *Dot Pattern:* Tests the monitor's capability to display images created by a montage of multi-colored dots.

- *Vertical Bars:* Draws patterns created primarily by bars running up and down the screen.

- *Horizontal Bars:* Draws patterns created primarily by bars running across the screen.

- *Text Color Chart:* Displays the different colors in which text can be displayed.

- *Verify Screen Mem(ory):* Tests the memory in an address space reserved for the information to be sent to the video card.

- *Character Set:* Verifies that characters from the standard ACSII character set can be displayed.

- *Video Attributes:* Tests nontext characteristics of the display, such as brightness, color rendition, flash accuracy, and intensity.

- *Fill Display:* The entire screen is filled with characters, testing its capability to compose and then properly display these characters. It also tests the screen's capability to autowrap text.

- *Cursor Addressing:* Translates the addressing sequence that identifies where the screen cursor should appear and confirms that the cursor appears at the correct location.

- *Video Memory:* Sends multiple sets of data to memory installed on the video adapter and confirms that identical data is provided by that memory when read back.

- *Hi-Res Color:* Tests color rendition at higher resolutions than standard VGA.

- *Text Resolution:* Tests the reproduction text at various resolutions.

- *Parallel Port (1, 2, and 3):* Performs an I/O operation across the parallel ports and reports back the I/O address of that port. Requires that a loopback adapter be installed on the port being tested.

- *Serial Port (1, 2, 3, and 4):* Performs an I/O operation across the serial ports and reports back the I/O address of that port. Requires that a loopback adapter be installed on the port being tested.

- *Reset Drive:* Tests the integrity of the Change Drive signal that informs the controller that the disk in the drive has been changed.

- *Rotational Timing:* Confirms that the R/W heads are in synch with the rotational speed of the motor.

- *Seek/Read:* Tests the drive's capability to locate a specific piece of data and read it back to memory.

- *Read Only:* Performs a read operation from a preselected sector and compares the data it finds to that which the test is expecting.

- *Controller:* Tests the capability of the hard disk's controller to queue and execute commands properly.

- *Seek (Hysteresis):* Performs a series of I/O operations requiring that the data be located and tracked by the controller.

- *Surface Test:* Examines each cluster on the drive looking for damaged or weak sectors.

- *All Zeros:* Fills all available memory with 0s and then reads it back, looking for the presence of any 1s.

- *All Ones:* Fills all available memory with 1s and then reads it back, looking for the presence of any 0s.

- *Checkerboard:* Alternates 0s and 1s across the memory matrix in a checkerboard pattern, and then reads it back, looking for any deviations in the pattern.

- *Address:* Writes data to each available memory address and then performs an I/O operation to that address to verify that the data written is found at the proper address.

- *Walking Ones:* Takes a pattern consisting of several contiguous 1s, while the remainder of free memory is written to 0s, and moves the block of 1s in increments equaling the number of 1s in that block up through the remainder of available memory. In a system with a large amount of RAM, this test can take a long time.

EXERCISE 1 REVIEW

As you can see, troubleshooting doesn't have to be a trial-and-error sort of thing. With the proper tools at your disposal, a lot of the guesswork can be eliminated.

1. Why can't you run TuffTEST from your Technician's Boot Disk?

2. Why isn't TuffTEST able to test certain devices, such as modems and sound cards? (Be careful here, you might have to think back to material presented in the textbook to answer this one)!

3. As noted in the lab, in order to test either a serial port or a parallel port using TuffTEST (or most other diagnostics programs, for that matter) a loopback adapter must be installed. Why do you think this is the case?

EXERCISE 2: WORKING WITH A POST CARD (OPTIONAL)

A POST Card (**Figure L10.3** and **Figure L10.4**) is a hardware diagnostics tool that reads BIOS calls throughout the process of POST. You insert the card into an available expansion slot (usually PCI, but some of the older ones are ISA) and turn on the computer. An LED display flashes result codes for each POST operation until such time as the boot process fails. At that point, the error number displayed shows what process was active when POST failed, and, therefore, what component failed. To perform this exercise, you will need a POST Card and the accompanying diagnostics manual along with a computer that can't complete POST. You can create the latter by removing the memory or video card, but without the manual, you won't be able to interpret the codes.

1. To come up with an unbootable machine, remove either the memory or the video card.

2. Examine the motherboard to determine the brand and version of BIOS installed.

3. Insert the PCI POST Card into any free expansion slot of the appropriate type.

4. Power the machine up.

5. Record the last code displayed. Depending on the brand, this code may be displayed in either decimal or hexadecimal notation.

6. Look up the code in the manual. It will tell you where the failure occurred.

Figure L10.3 The PCI POST Card doesn't look much different than any other expansion card, except for the little LED display in the upper right-hand corner.

EXERCISE 2 REVIEW

1. What does the final code glowing on a POST card represent?

2. On any of the POST card messages that points to a chipset failure, what has the technician learned?

LAB SUMMARY

Until now, you might have thought that all hardware troubleshooting was the result of the exquisitely trained mind of the technician coupled with finely-tuned intuition. Now you know better; you cheat.

Figure L10.4 When inserted, it's easy for the LED display to be blocked by other devices.

   No technician is going far in this industry without a good set of diagnostics tools and the skills with which to use them. Properly armed, the technician looks very impressive indeed.

Working with SCSI

For years now, the computing industry has accepted the fact that high-end workstations and servers need to be equipped with SCSI. The advent of technologies such as Serial ATA has slowed this tendency in workstations, but servers, for the most part, still make use of SCSI.

Newer implementations of SCSI, such as Fiber Channel Arbitrated Loop, are generally trouble-free and fairly automatic in their installation. Also, FCAL is usually installed and configured at the factory on most servers. However, older technologies such as Parallel SCSI still require a bit of knowledge on the part of the technician. Therefore, in these exercises, you will be installing and configuring a SCSI-II chain.

To complete these exercises, you will need the following at the minimum:

- The student computers, equipped with keyboard, monitor, and mouse
- A PCI SCSI host adapter
- A SCSI cable with a minimum of three device connectors, in addition to the connector for the controller
- A SCSI hard disk (two if possible)
- Any other secondary internal SCSI device (a tape drive or CD-ROM will be suitable)
- The manuals for the SCSI devices and controller

A few CompTIA objectives will be made clearer in these exercises as well. Among them are the following:

1.2 Identify basic procedures for adding and removing field-replaceable modules for desktop systems.

1.3 Identify basic procedures for adding and removing field-replaceable modules for portable systems.

1.4 Identify typical IRQs, DMAs, and I/O addresses and procedures for altering these settings when installing and configuring devices. Choose the appropriate installation or configuration steps in a given scenario.

1.6 Identify proper procedures for installing and configuring common IDE devices. Choose the appropriate installation or configuration sequences in given scenarios. Recognize the associated cables.

In the following exercises, I will be working with 50-pin SCSI. If you purchased your components new, they are more likely to be either 68- or 80-pin devices and cables.

EXERCISE 1A: INSTALLING THE HOST ADAPTER

The first thing that you need to do is install the host adapter (**Figure L11.1**). For these labs, I have specified that the classrooms be equipped with SCSI expansion cards. However, note that many high-end motherboards are equipped with on-board SCSI, and this step would be required only in the event that a second SCSI chain was required.

Figure L11.1   A SCSI host adapter will be equipped with at least one internal connector for a SCSI cable. Most, such as this Adaptec 2940AU, are also equipped with a connector for hooking up external devices.

1. Examine the manual for the host adapter and see whether configuration settings are done via a setup program or whether jumpers or dipswitches are involved. A Plug 'n Play host adapter will most likely not have any manual settings for IRQ and/or I/O channel. However, manual settings may exist for terminating the adapter and for setting its device ID. On most (if not all) newer devices, even these settings are accomplished in the setup program. If jumpers are to be set, now is the time to set them.

2. Open the case and locate a free expansion slot.

3. Carefully insert the host adapter into the slot and tighten the screw that affixes the adapter to the backplane.

4. Attach the SCSI cable (**Figure L11.2**) to the internal connector. If you are using two-channel adapters, it is best to use Channel A. You're now ready to tackle the devices.

Figure L11.2   This particular cable is designed for four devices in addition to the host adapter. This is a deluxe cable in two respects. First, each connector is equipped with a pull loop for removing it from the device later. Second, if you notice on the end of the cable, the cable itself is terminated, so you don't have to worry about terminating individual devices. Just make sure they're *not* terminated.

EXERCISE 1A REVIEW

1.   How is installing a SCSI host adapter similar to installing any other device?

2.   What is the default IRQ for the SCSI host adapter?

EXERCISE 1B: INSTALLING THE SCSI DEVICES

In the physical installation of SCSI devices, not that many differences exist between those devices and the IDE devices you installed in Lab Four, Installing IDE Devices. Some critical differences do exist in configuration and cabling.

1.   Examine the manual for the device to see how device IDs are set for the particular device. The Seagate Barracuda seen in **Figure L11.3** has a series of jumpers. As the illustration indicates, the smaller block is for terminating the drive, and the larger one is for various other settings. As far as other models of drive are concerned, the variety of options is as wide as the variety of devices. That's why you need the manual.

2.   Set each of the devices that you're installing to a different device ID. If using a terminated SCSI cable, such as the one in Figure L11.2, all devices should be unterminated. If the cable is not terminated, the last device on the chain needs to be terminated. Again, the various methods for terminating devices vary. In the card illustrated in Figure L11.3, jumpers are used. Other methods include a terminating resistor that is plugged into a socket and software-configurable termination. In the latter case, the device is completely configured in the setup program.

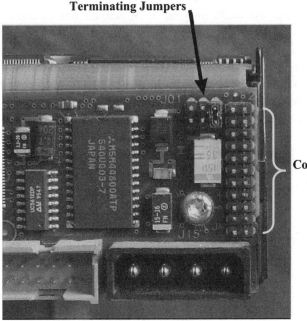

Terminating Jumpers

Configuration Jumpers

Configuration jumpers are used for setting device IDs and for other settings, including parity, motor startup delay and others

Figure L11.3   On this Seagate Barracuda, device IDs and other settings are made manually with these jumpers.

3. After the device ID and termination are properly configured, install the devices in the appropriate drive bay and attach the cables. The Molex cable is no different than that of the IDE devices you installed. However, unlike IDE devices, plugging a SCSI cable in backward can damage or destroy the device and/or the host adapter. On the 68- and 80-pin cables, this is not an issue, because both cables use D-shell connectors. They only plug in one way. On the 50-conductor cables and connectors, most are keyed to prevent this from happening, but not all. If the cable is not keyed, make sure that the colored conductor on the ribbon cable is lined up to pin number 1.

4. Another thing to consider is that most computers have an LED disk activity light on the front panel. If you disconnect the cable from the LED connector on the motherboard and connect it to the LED connector on the host adapter, the LED on the front panel of the computer will light whenever there is activity on the SCSI bus. That way, regardless of which hard drive is active, the LED will flicker.

Exercise 1b Review

1. Why is it so necessary to terminate the SCSI chain?

2. What are two common methods of terminating a SCSI chain?

Exercise 1c: SCSI Setup

After the devices are installed, it's time to check your setup. When you first turn on the system after installing a SCSI host adapter, if the adapter is successfully recognized by the system, you'll see some changes in your POST. Immediately after the conventional POST to which you grew accustomed, a message will appear telling you that the BIOS for your SCSI adapter was successfully loaded. You will also be prompted to press a specific key to enter Setup. This is not the same setup you looked at in Lab Nine, A Tour of the BIOS. This is the SCSI Setup.

Immediately after this prompt, the SCSI host adapter will scan the SCSI bus looking for all installed devices. If you have properly configured the devices you installed, each on a different device ID, then the adapter should have no trouble identifying the devices and reporting the device IDs that they have claimed. In this case, there is really no reason to enter the SCSI setup. But what fun would that be, and how could I lead you on a tour of what you can accomplish in the setup program?

Because different folks will be using different setup programs, I am not going to clutter this section with screen shots that might be confusing to many. Start your setup program and look for the following settings. Note that not all setup programs contain the same settings, but most should have the ones discussed here.

1. SCSI Bus Interface: Enables the user to configure IRQ and determine whether or not to use parity checking.

2. Boot Options: Determines whether or not to boot from the SCSI bus, and if so, which device ID identifies the boot disk. On many brands, this is also where Logical Unit Numbers (LUN) are configured for multiple devices on the same device ID. This will tell the host adapter to treat all devices with the same device ID, but with unique LUNs, as a single device.

3. SCSI Device Configuration: Enables the user to configure maximum transfer rate for a specific device, enable write-back cache, and tell the host adapter whether the device can negotiate bus

width (such as Wide SCSI or Fast/Wide SCSI). On many brands, there will be an additional setting for configuring LUNs. This setting must be enabled for multiple devices to span multiple LUNs on a single device ID. The default setting is usually disabled. Also, in many brands, you can dictate whether or not you want the SCSI scan to be a part of your BIOS scan. Unless you have an overpowering reason for not doing this, I recommend that you leave this setting at the default, which is enabled.

4. Advanced Configuration Options: Not all adapters will have these settings.

    a. SCAM Support: Newer adapters make use of a protocol known as SCSI Configuration Automatically (SCAM). If enabled, this gives Plug 'n Play capability to the SCSI bus. If the devices also support SCAM, then the host adapter automatically assigns device IDs.

    b. Reset SCSI Bus at IC Initialization: During startup, or after a hard reset, the host adapter will automatically issue a RESET command to all devices on the bus.

    c. Support Removable Disks Under BIOS as Fixed Disks: Provides three options.

       i. Boot Only: A removable media drive that has been designated as a boot device will be treated as though it were a hard disk.

      ii. All Disks: All removable media will be treated as though it were a hard disk.

     iii. Disabled: No removable media will be treated as though it were a hard disk.

    d. Display <Ctrl-A> Message During BIOS Initialization: This message may vary from brand to brand, but it lets the user determine whether or not the SCSI BIOS will issue a prompt as to what key sequence to press in order to enter SCSI BIOS Setup.

There are most likely other settings visible as well, but these are the ones that are universally important.

EXERCISE 1C REVIEW

1. What is the difference between a device ID number and a Logical Unit Number?

2. What happens with the device ID of three devices are all the same, when each one has a different LUN?

LAB REVIEW

1. What different settings might have to be manually configured on an older host adapter?

2. What is one critical difference between hooking the ribbon cable up to a SCSI device and hooking one up to an IDE device?

3. You've just now installed a new SCSI chain, and nothing works. During POST and in Windows Device Manager, the host adapter is recognized. What is a likely reason for failure?

4. You've just installed a new device on an existing SCSI chain. All the old devices still work, but the new device isn't recognized. What are two things that might cause this problem?

5. How would you configure three different hard drives to work as a single array?

LAB SUMMARY

If you think this process was bad, you should have seen what technicians went through in the old days of SCSI. Back then, it wasn't uncommon for the host adapter from one manufacturer to refuse to recognized devices from another. And all the settings were manually configured! As time goes on, it's going to get even easier. Most serial SCSI devices are auto-configuring and require little or no attention from the technician. Still, when it comes time to configure that RAID array, it's nice to know that you have some idea of what you're doing.

MAKING CABLES

One of the more common tasks that befall a hardware technician is that of preparing cabling and/or wiring patch panels. Neither one of these tasks results in much mental stress, but they do call on your physical dexterity a bit.

For these exercises, you're going to need the following items:

- Network cable crimping tool

- Bulk CAT5 or CAT5e twisted pair cable

- RJ-45 terminators

- 18- or 24-port patch panel

- Punchdown tool

By the time you're finished, you will have prepared the patch cable that you will use to connect your computer to the network. You also will have made a crossover cable that you can add to your toolkit, and you will have learned how to wire a patch panel in a wiring or server closet. Do whatever warm-ups you choose for fine-tuning your coordination, and let's go. Some CompTIA exam objectives you'll see along the way include the following:

1.5   Identify the names, purposes, and performance characteristics of standardized/common peripheral ports, associated cabling, and connectors. Recognize ports, cabling, and connectors by sight.

1.6   Identify proper procedures for installing and configuring common IDE devices. Choose the appropriate installation or configuration sequences in given scenarios. Recognize the associated cables.

EXERCISE 1: A TOUR OF THE CRIMPING TOOL

Pick up and examine your crimping tool. You'll notice three key functions to this tool. **Figure L12.1** will help you find these features if you're new to this tool.

There is a wire-cutting blade, which, by the way, is used for cutting the lengths of cable. You'll notice that the second set of blades, closer to the end, does not close all the way. These blades are for stripping the outside layer of insulation from the cable. Be careful using this feature. With many of these tools, it is very easy to snip off one or more of the wires inside. You don't want that. Last is the crimping jaw itself. Conveniently shaped like an RJ-45 jack, simply bring the handles together, insert the terminator, and squeeze.

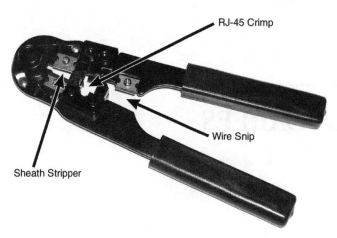

Figure L12.1   The RJ-45 Crimping Tool

EXERCISE 1 REVIEW

1. Carefully measure the depth of the cavity into which the cable seats for stripping off the outside insulation. How much insulation does it strip off?

2. Does the wire-stripping tool close all the way?

EXERCISE 2: MAKING A PATCH CABLE

You will need a length of CAT5 or CAT5e twisted-pair cable of the appropriate length to connect your system to the wall jack in your room. You'll need an appropriate length of CAT5 or CAT5e cable, two RJ-45 (RJ-45 stands for *registered Jack 45*) terminators for the patch cable (but it would probably be a good idea to have extras lying around, just in case), and your own crimping tool. This will go much faster if everyone has his or her own crimping tool.

Strip approximately 3/4 of an inch to an inch of the outer insulation from each end of the cabling, exposing the four pairs of wire. Do *not* attempt to strip each individual wire.

> **NOTE:** Most crimping tools have a spacer on the sheath stripper that trims off the correct length for a standard RJ-45 connector. Personally, I find it easier to arrange the wires into the correct order if I strip an inch or more. After the wires are arranged, flatten them between your fingers and use the wire snips to trim them evenly to the desired length.

Note that the wires are arranged in pairs of wires twisted together (hence the name twisted pair). These pairs are green-white/green, brown-white/brown, orange-white/orange, and blue-white/blue. Untwist the pairs and arrange them into the following order: white/orange, orange, white/green, blue, white/blue, green, white/brown, and brown (as in **Figure L12.2**).

Making sure that the wires stay in the correct order, press them snugly into the connector. With the locking clip facing away from you, the orange pair should be on the left (**Figure L12.3**.)

Place the connector into the jaws of the RJ-45 crimp and squeeze the handle firmly together as in **Figure L12.4**.

Repeat the process on the other end of your cable, and you have a finished patch cable.

EXERCISE 2 REVIEW

1. What would be the result of getting two wires crossed when seating them into the connector?

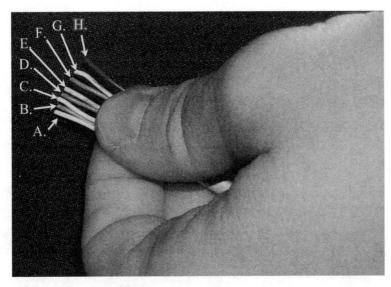

A.   White/Orange

B.   Orange

C.   White/Green

D.   Blue

E.   White/Blue

F.   Green

G.   White/Brown

H.   Brown

Figure L12.2   The Proper Order of Colors When Making a Patch Cable

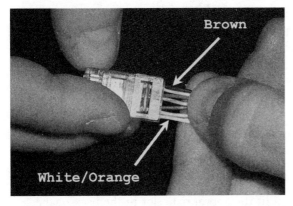

Figure L12.3   Positioning Your Cable in the Terminal

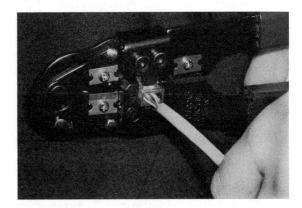

Figure L12.4   Crimping the Cable

2.   List the order of wires as they go into the connector.

3.   What does RJ signify?

EXERCISE 3: MAKING A CROSSOVER CABLE

A crossover cable is a handy accessory to have in your gadget bag. It enables you to directly connect two PCs, a notebook to a PC, or any other two devices that normally make use of straight-through RJ-45 interfaces. To make one, repeat the preceding procedures with this exception.

One end of the cable will be wired as described in Exercise 2. The other end will be wired in the following order: white/green, green, white/orange, blue, white/blue, orange, white/brown, and brown.

EXERCISE 3 REVIEW

1.   What specifically is the difference between wiring a standard patch cable and wiring a crossover cable?

2.   How is a crossover cable useful?

EXERCISE 4: WIRING A PATCH PANEL

Most networks these days bring the cable drops together into a single location and connect them to a patch panel. The patch panel then is labeled as to what device is attached to what port. That way it is easier to disconnect a specific device from the network or to patch it into a different subnet if necessary. To wire a patch panel, you need the patch panel itself, cable drops, and a punchdown tool.

Punchdown tools run the gamut from a simple plastic tool to fancy electrical devices that pound the wire into the panel with enough force to knock out an elephant. The small gray device shown in **Figure L12.5** was bundled with an RJ-45 wall jack. It's the one I find myself using all the time, because carrying the other one around in my pocket seems like a risky proposition to me.

Patch panels generally ship with colored guides to indicate the proper wiring pattern. Note that there are two different standards. The Electronics Industry Association teamed up with the Telecommunications Industry Association to form the EIA/TIA 568A and 568B standards. Make sure you use 568B.

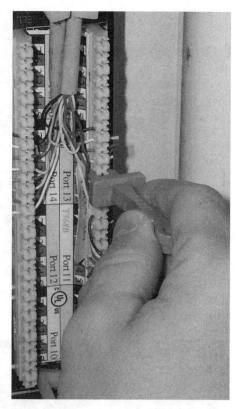

Figure L12.5   Punchdown tool

1. Pull down the cable drop you want to wire, run it through the cable channels in the panel, and clip it off to the length you need.

2. Strip off about two inches of the insulation sheath and separate the wire pairs.

3. Trim each individual wire to the correct length to fit the appropriate slot in the back of the patch panel.

4. Line the wire up to the correct slot and press it down firmly with the punchdown tool. Repeat the process until all eight wires are seated.

EXERCISE 4 REVIEW

1. What makes a patch panel such a convenient device to add to any network?

2. Why wouldn't you want to wire your patch panel to EIA/TIA 568A standards?

LAB SUMMARY

Just having an A+ certification is no guarantee that you'll ever be called upon for any networking chores. On the other hand, it's no guarantee that you won't, either. Any time you find yourself in the position of being considered a professional in the information technology industry, you need to be prepared for simple tasks and be able to make patch cables with your your eyes closed.

SETTING UP A PEER-TO-PEER NETWORK

In the following exercises, you'll get your first taste of actually networking multiple devices. You will interconnect the computers onto which you installed Windows 98 in Lab Six and hook them all together in a small peer-to-peer (P2P) network. But before you do that, you're going to need NICs installed. To complete this lab, you will need several items:

■ The student machines

■ A NIC for each machine, with drivers (if necessary)

■ Tool kits

■ A hub

■ The patch cables you made in Lab Twelve

The sole CompTIA exam objective that you will encounter in this lab is as follows:

6.1   Identify the common types of network cables, their characteristics, and connectors.

EXERCISE 1: INSTALLING A NIC

Place the computer on the desk in front on you. Depending on the model, you will either need to remove the case or the access panel in order to add components. Because the methods for doing this vary greatly from one model of computer to another, it isn't practical to provide instructions on how to open the computer in this text. If you are unclear on how to open your computer, consult with your instructor.

Now it's time to read any instructions that come with your NIC. Some brands, such as 3-Com, require that you run a small setup program before installing the NIC, and others have you install the driver after the card is in place.

Find a free PCI slot and remove the backplane cover for that slot (see **Figure L13.1**). Firmly seat the NIC into the slot and replace the backplane screw.

Start the computer and let Plug 'n Play do its thing. With Windows 98, the NIC should be detected automatically. Depending on your card and the installation process used, it will either identify the card and find the appropriate driver in its database, or it won't. If you ran a setup program such as 3-Com's prior to installing the card, it added its drivers to the database. If no driver is found, during the

Backplane cover

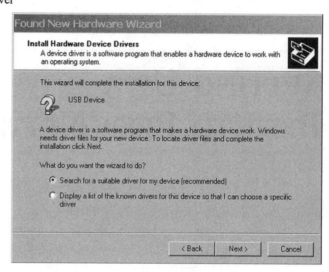

Figure L13.1 The backplane cover (raised in this illustration) is actually an essential part of your computer's thermal regulation system. Any time you remove any card, fill the open slot with a cover.

Figure L13.2 In most circumstances, it is best to let Windows search for a device driver. Select that option for this lab.

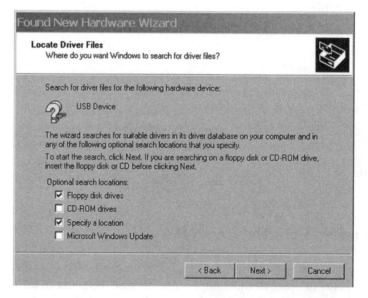

Figure L13.3 Click the appropriate box that specifies where your driver is located. To save a little time, deselect everything else.

installation process you will get a screen like the one in **Figure L13.2**, followed by the one in **Figure L13.3**. Specify the appropriate location of the disk containing your drivers and click Next.

Your NIC installation is complete.

EXERCISE 1 REVIEW

1. When installing a NIC, what are two different ways device drivers might be added?

2. You have a NIC but no device drivers. Can you make it work?

EXERCISE 2: CONFIGURING THE PROTOCOLS

After installing a NIC driver, Windows automatically installs certain networking functions. By default, Windows installs a networking client in Client for Microsoft Networks. It also installs File and Printer Sharing for Microsoft Networks and Internet Protocol (TCP/IP). TCP/IP is configured to automatically obtain an IP address from a DHCP server. Therefore, your machine is configured as a DHCP client as well. You're going to change some of that.

EXERCISE 2A: RECONFIGURING TCP/IP

The first thing you'll do is statically assign IP addresses to each machine on the network.

1. Right-click My Network Places on the desktop and click Properties, or click Start → Settings → Control Panel → Network and Dial-up Connections. Either way, you will get to the screen in **Figure L13.4**.

2. Scroll down to the TCP/IP for your NIC. In my computer, it is the SIS 900 Fast Ethernet Adapter. In yours, it will be whatever brand is installed. Click properties. The screen in **Figure L13.5** should appear.

3. You will statically configure each of the machines in the classroom. To do this, click Use the following IP address: and type the address assigned to your machine. You will derive the address by starting in the front left corner of the classroom and moving left to right and then to the back, one row at a time. Start with the IP address of 192.168.1.100. The next machine will be 192.168.1.101, followed by 192.168.1.102, and so on. Make sure that no two machines are inadvertently assigned duplicate addresses. This will cause a conflict. Type a subnet mask of 255.255.255.0. If you press Tab after entering your IP address, this field should fill in automatically. Click OK, and then on the next screen, click Close.

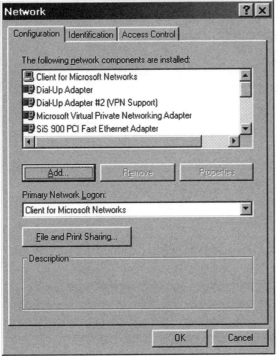

Figure L13.4 Network and Dial-up Connections

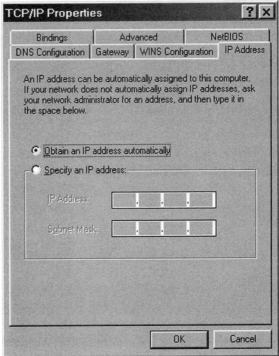

Figure L13.5 TCP/IP Properties

EXERCISE 2B: CREATING A WORKGROUP

Now click the Identification tab. You should get the illustration shown in **Figure L13.6**. Make sure Workgroup is selected and type **CLASSROOM** in the Computer Description field. Everybody needs to be in the same workgroup to be part of a P2P network. Therefore, if you type it incorrectly, you won't join the network.

Congratulations. You're now all part of the CLASSROOM workgroup. You should be able to see the CLASSROOM workgroup in My Network Places. If you double-click this icon, you should be able to see each other's machines. Note, however, that you won't be able to see any of the contents of these machines as of yet.

EXERCISE 2C: SHARING RESOURCES

1. The first thing that you must do is make sure that File and Printer Sharing is installed as a service on your computer. To do

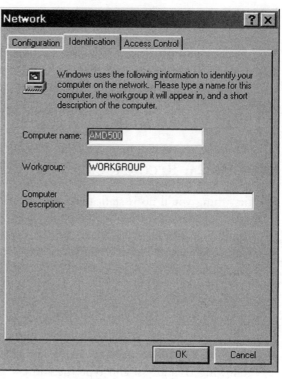

Figure L13.6   Configuring the Workgroup

that, open the Network applet once again and click the File and Printer Sharing button. Check the box next to I want to be able to give others access to my files.

2. Reboot the computer.

3. Right-click the Start button and select Explore. This will open Windows Explorer, as shown in **Figure L13.7**. You're going to create a new folder in which you can store your shared files.

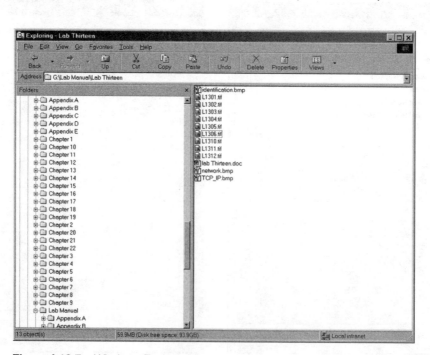

Figure L13.7   Windows Explorer

4. Highlight Drive C in the left-hand pane. Select File, New, and then Folder. Give the folder your first name.

5. When the new folder appears in the left-hand pane, right-click that folder, and in the pull-down menu that appears, one of the options will be Sharing. Click that option. The screen in **Figure L13.8** will appear.

6. Select Share this folder and leave the default share name in place. Don't worry if more than one of you have the same first name. Just don't try to put two shared folders with the same shared name on the same computer. You won't worry about permissions in this lab, so click Apply and then OK.

7. Now, in the left pane of Windows Explorer, double-click your shared folder. The right-hand pane should be empty. Right-click anywhere in that pane, and you'll get a pop-up menu. Select New, and then in the next menu that appears, select Text Document.

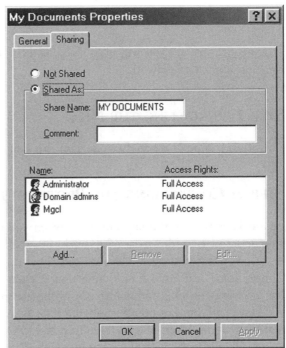

Figure L13.8   Sharing a File in Windows 98

8. Type the following text. "My house is brown, but my horse is yellow." Don't worry about how stupid that sounds. One of your classmates is going to fix that for you. Close the document and, when it prompts you to save the file, save it as house.doc.

9. Now everybody pick a teammate. Pick somebody you don't usually socialize with so that you can learn to work together and maybe make a new friend. In Windows Explorer, double-click My Network Places and browse to your partner's machine. Open the file and edit it to read, "My house is yellow, but my horse is brown." Save the file and close it. If your new shares don't appear immediately, don't panic. Sometimes it takes a few minutes for the Windows Browsing Services to advertise the new shares.

10. When you're both finished, you should be able to open your version of house.doc and see the revisions your partner made.

EXERCISE 2 REVIEW

1. What are two different ways of getting to the Network Applet in Windows 98?

2. What are two settings that must be configured if you're statically assigning an IP address to a NIC?

3. In a Workgroup setting, two configuration settings in the Identification tab are required. What are they, and what are the rules concerning them?

4. How do you share resources in Windows 98?

5. Why might a new share not immediately appear in Windows Explorer on remote computers?

OPTIONAL EXERCISE 3: DUAL-HOMING A COMPUTER

To perform this lab, you will need a second NIC for each computer. Each NIC will be configured to work on a separate network.

STEP 1: INSTALLING THE NIC

This is easy. Simply repeat Exercise 1 of this lab, step by step.

STEP 2: CONFIGURING THE PROTOCOLS

Repeat Exercise 2 of this lab. However, when you get to Procedure 5 of Exercise 2a, instead of starting your series of IP addresses at 192.168.1.100, start with 192.168.10.100.

Now when you open Network and Dial-up Connections, you will see an icon for Local Area Connection 1 and Local Area Connection 2. Each one can be independently configured. You now have two networks running simultaneously in the classroom.

Have half the classroom reconfigure their Network Identification in the System Properties screen so that they are members of a workgroup called WORKGROUP. You will have to restart these machines for the change to take effect. In My Network Places, you should be able to browse to both workgroups.

LAB REVIEW

1. List the different steps involved in installing a NIC.

2. How would you go about configuring a workstation to use static IP addresses in TCP/IP?

3. Go over the steps involved in setting up a peer-to-peer network, from beginning to end.

4. Why might you want to dual-home a particular computer?

5. What is different about dual-homing?

LAB SUMMARY

As you can see, setting up a peer-to-peer network isn't all that complicated. Makes you wonder why you didn't network all twelve of your computers a long time ago, doesn't it? The key here is paying attention to the details. It's the little things that will get you, like having ten of twelve computers part of the SALESTEAM workgroup and the other two in the SALETEAM workgroup. (I hate it when that happens.)

BUILDING A VIDEO WORKSTATION

One of the trends I see with a vast number of computer hardware technicians is this. I ask them to look on the front bezel of their own personal computers to see what brand they prefer. And what brand do they find as often as not? No brand at all. Rather than trust a system put together from parts supplied by the lowest bidder, many technicians choose to purchase the parts they need from reliable vendors and assemble their own computers. This enables the technician to pick and choose each component that goes into the system based on his or her own set of priorities.

The way those priorities are established is based on a very simple question. What is the primary use to which this system will be put? I discussed this in great detail throughout the textbook. Different applications have different requirements. For the person who needs nothing more complicated than an intelligent typewriter (Professor, what's a typewriter?), there isn't much need for a system with huge amounts of horsepower. However, the avid gamer and the digital photographer have discovered that the more computer horsepower the system has, the easier their work is. Well, faster, anyway.

In this exercise, you're going to build (on paper, anyway) a high-end video workstation designed to create full-length animated video presentations along the lines of those put out by the famed Pixar Studios. You may recall some of their classic titles such as *A Bug's Life*, in which the 3D graphics renditions are so lifelike you almost want to reach out and touch them.

Now, I realize that a number of you who are more experienced will complain that the PC platform is not suitable for anything this advanced. And it is true that, for the majority of professional studios involved in this type of work, the PC is somewhat scorned. I'm here to tell you that it can—and *will*— be done, if you want to get a passing grade out of this course. Then again, I guess that's up to your instructor.

This is an exercise in theory, rather than reality. So the only materials you need are a good imagination, a pencil and paper (a word processor will do if that's all you can come up with), and a viable Internet connection. I strongly recommend that this and the following lab be introduced very early in the semester so that students can be thinking about their configurations as they are introduced to the various components and their functions as the course progresses.

As I said, this is an exercise in theory, and the students are not expected to actually purchase the parts and build the system. However, should you choose to do so and would like to have your system evaluated, send me an email and I'll provide you with a shipping address. I'll be happy to put your finished product through its paces for a year or so, and then I'll get it right back to you.

You're going to be putting the CompTIA objectives through their paces while doing this exercise. Among the objectives you'll cover are the following:

1.1 Identify the names, purpose, and characteristics of system modules. Recognize these modules by sight or definition.

1.2 Identify basic procedures for adding and removing field-replaceable modules for desktop systems.

1.3 Identify basic procedures for adding and removing field-replaceable modules for portable systems.

1.8 Identify proper procedures for installing and configuring common peripheral devices. Choose the appropriate installation or configuration sequences in given scenarios.

4.1 Distinguish among the popular CPU chips in terms of their basic characteristics.

4.2 Identify the types of RAM (Random Access Memory), their form factors and operational characteristics. Determine banking and speed requirements under given scenarios.

4.3 Identify the most popular types of motherboards, their components, and their architecture (bus structures).

Building the Video Workstation

The dream system that you're aspiring to create must be capable of several different, and in many cases, very complex, tasks. You are going to create a streaming audio/video presentation of a minimum of one hour in length. In addition to producing this project, the system you design must be able to play it back. Now that last part probably sounds pretty easy, but I'm going to throw in a catch. It must not only play it back in compressed MPEG format; it must be able to play it back in its raw file formats as they are originally produced in editable form. On top of this, I don't want the presentation being shown on the computer monitor. I want to display it in optimum quality to a room of twenty people. One of the decisions you're going to have to make in designing your system is what method you will use to project your finished product to an audience. Then you're going to create a Website onto which you will post a short sample of your presentation.

To properly provide a reference to the performance required, it is only fair that I provide a list of software applications that must run on this system for it to be considered a successful design. The following list does not constitute an endorsement for these products. They are all very fine products, but I'm sure others are equally good, and perhaps some are even better. I selected this list of applications for one very scientific reason. I've used them all at one time or the other, so I'm very familiar with them.

- Adobe Photoshop
- Adobe Premier
- Adobe After Effects
- Adobe Audition
- Dreamweaver MX
- Microsoft Office 2003

Performance Considerations

Now, I'll be the first to admit that a large number of off-the-shelf computer systems sold by the vast majority of major manufacturers are perfectly capable of running all of the preceding applications. The key to this assignment is that you're designing your own system as a way of getting around the performance compromises inherent in any of the pre-fab machines.

Processing

A couple of the applications in your list are extremely processor intensive. Therefore, you need to design your system so that the processor does not become a bottleneck. Refer to Chapter Seven of the textbook, Understanding CPUs, for a more detailed discussion of different processor features to consider in this respect.

Memory

Nearly all of the listed applications are hungry for as much memory as you can throw at them. But this is definitely an area where sheer quantity will not be your only ally. You must not only design your system with sufficient memory for the task at hand, but with the appropriate type of memory. Review Chapter Nine of the textbook, Searching Your Memory, before committing to a decision.

I/O Bus

Most of the applications you'll be running will be moving huge quantities of data back and forth across the different system busses. The various components you choose for your system will need to attach to these busses. It won't matter much if you've picked the perfect CPU/memory combination if you're trying to move all that data over an 8.33Mhz ISA video card. But you won't be trying that; will you? Perhaps a second reading of Chapter Ten of the textbook, Examining the Expansion Bus, will help.

Graphics

Here's where the gamers among you might have to think a little bit outside the box. Sure, you're already up to date on all the fastest gaming cards and what makes them tick. But is speed all you need here? The audience won't be all that impressed with the speed if they can't see it. And they *really* won't be impressed if, for some reason or the other, the video card you chose won't even play back the format you chose. You might want to brush up on computer video a bit by re-reading Chapter Sixteen of the textbook, Your Computer and Graphics.

Hard Disk Storage

These days, the size of your drive isn't what matters. It's how you use it. The smallest hard disk manufactured today will provide sufficient space for all of your applications and still leave room for several projects like the one described in this exercise. However, when you go to play back that 60fps hi-res masterpiece, it's really going to hurt your pride and your professional image if it looks all jittery. Read Chapter Fourteen of the textbook, Hard Disk Interfacing and Disk Management, and perhaps Chapter Fifteen, The Many Faces of SCSI, before you tackle this decision.

Alternative Storage

When you've completed your project and had your twenty reviewers take a look at it and provide their critiques, you're going to need to do one of two things. Go back to your day job, or package up your masterpiece for sending out to a professional studio for distribution. You're hoping for the latter. So you're going to need a storage medium that is mutually acceptable to you and the studio of your choice.

To figure out which medium you should use, refer to Chapter Seventeen of the textbook, *Multimedia: Computerized Home Entertainment.*

THE ENCLOSURE

Now you need to find a home for all these fancy new components. It probably won't be such a great idea to spread them out over the counter and run wires and cables back and forth between them. Also, what about all those drives you've selected? What do you do with them? You need a computer case. But what kind of case do you need, and just where are you going to put it?

ULTERIOR CONSIDERATIONS

Now here's the real kicker. In reality, you aren't the production engineer in our little scenario. You are a sales representative for a company that specializes in custom configurations, and you're designing this machine for a client. Besides these performance demands that you must meet, you're in competition with several other companies. These companies are your other classmates. For this project, only one A will be assigned. And that will be given to the student with the computer that most successfully meets the performance demands, but at the lowest price. Your instructor is the purchasing agent whom you must please, and his or her final judgment is all that matters.

BUILDING A NETWORK SERVER

In the previous lab, I had you design the perfect video workstation. That was fun, wasn't it? Although I'm a little disappointed that UPS still hasn't delivered any examples of your work for me to test yet.

Now you're going to build a server for a network. As with the previous exercise, you will be acting out the role of a technical salesperson for a manufacturer and your instructor will be the purchasing agent. The best configuration for the lowest price wins.

Once again, your priorities will be based on the same question put forth in the last exercise. What is the primary use to which this system will be put? However, the applications running on a typical server are a bit different. And you also have to take into consideration the impact that the performance of various components will have on network users.

In this exercise, I'm going to have you assemble a network server that will handle user logons and an email server and also act as a file server for a network of 100 users. The files stored on the server range from word processing documents that are under 100KB to an Oracle database that is a whopping 62GB. On top of this, you're going to use your server as an application server!

Once again, here come the whoops and hollers from the more experienced geeks in the class. No Information Technology (IT) professional in his or her right mind would lay all of those responsibilities onto a single server. And those geeks would be right. However, in this case, the CEO has made the final decision and when it comes to money, everybody knows that there isn't a CEO out there with a mind. Don't even pretend that it's a "right mind."

As before, this is an exercise in theory, rather than reality. So the only materials you will need are a good imagination, your note taking equipment, and an Internet connection. And again, this lab is best served if it's introduced early in the semester so that the students can be thinking about their configurations as they are introduced to the various components and their functions as the course progresses. If you want me to test your server for you, you're going to have to get it to me a bit quicker than you have the video workstations so far.

The CompTIA objectives you're going to see are the same as the last lab, but since my editors insist that each lab be preceded by the objectives you'll encounter, here they are again.

1.1   Identify the names, purpose, and characteristics of system modules. Recognize these modules by sight or definition.

1.2   Identify basic procedures for adding and removing field-replaceable modules for desktop systems.

1.3   Identify basic procedures for adding and removing field-replaceable modules for portable systems.

1.8   Identify proper procedures for installing and configuring common peripheral devices. Choose the appropriate installation or configuration sequences in given scenarios.

4.1   Distinguish among the popular CPU chips in terms of their basic characteristics.

4.2   Identify the types of RAM (Random Access Memory), their form factors and operational characteristics. Determine banking and speed requirements under given scenarios.

4.3   Identify the most popular types of motherboards, their components, and their architecture (bus structures).

BUILDING THE NETWORK SERVER

A network server has a whole bunch of significantly different requirements than that of your video workstation. And while this isn't a class on networking, you should be familiar with the hardware that goes into servers as well as you are that which goes into workstations.

The applications that run on the server are a lot different as well, so you're going to need to research the requirements for the following products before beginning the search for components that will make these applications sing. Trust me when I tell you that this part may be as much work as researching the hardware you'll need. The applications you'll need to run are the following:

- Microsoft Windows 2000 Advanced Server
- Microsoft Systems Management Server
- Microsoft Exchange Server
- Oracle Application Server, version 10g
- Oracle Database, version 10g
- Veritas Backup Exec, version 8.6

PERFORMANCE CONSIDERATIONS

Servers are one component that IT professionals rarely, if ever, have built by a custom manufacturer. Generally, they entrust these purchases to the Dells, IBMs, Compaqs, and Gateways of the world. However, if they opened their eyes to just how much more inexpensively a server could be custom-built to their needs, they might change their minds.

PROCESSING

When building a server, the processor isn't just servicing the requests of the native applications. It also has to process logon requests from users just joining the network and the myriad of other services required by the users already on. How are you going to design your system so that one tiny little chip can do all that work? Or are you?

MEMORY

In one respect, this is like the previous lab in that all of the listed applications are hungry for as much memory as you can throw at them. However, if you've done your research on the software correctly, you're going to find out that the server needs memory for several other purposes. For example, how much RAM does each logged-on user require in Windows 2000 Advanced Server?

I/O Bus

The applications you'll be running on the server will still be moving huge quantities of data back and forth across the different system busses. But in this case, you might find that a completely different collection of peripherals might have priority over the others.

GRAPHICS

On the video workstation, the graphics subsystem was an integral part of the overall system. Its performance, as well as its overall capabilities, was critical to the overall mission. But just how much impact does the video system have on server performance? That's something you're going to have to decide.

HARD DISK STORAGE

Hard disk storage on the server requires a whole new way of thinking. Losing a project you've spent the last six months on is a tragedy. Losing all of a company's data that has accumulated over the past seven years is a complete and total disaster. But on top of that, 100 users are all trying to access that data at the same time. Oh, my! What are you going to do?

ALTERNATIVE STORAGE

You're probably not all that interested in sharing the data on your server with anyone from the outside world. But on the other hand, you might want to have some method of storing all the files that the Veritas Backup Exec is going to create. In fact, if Veritas doesn't find a device on which to store the data, it won't even install! So what kind of alternative storage should you use?

THE ENCLOSURE

Now you need to find a home for all these fancy new components. It probably won't be such a great idea to simply spread them out over the counter and run wires and cables back and forth between them. Also, what about all those drives you've selected? What do you do with them? You need a computer case. But what kind of case do you need, and just where are you going to put it?

ULTERIOR CONSIDERATIONS

Remember from the last lab? You aren't the production engineer in this little scenario. You are a sales representative for a company that specializes in custom configurations, and you're designing this machine for a client. Once again, you have a price point to meet, and once again you're in competition with your other classmates. As before, there will be only one A. And that will be the computer that most successfully meets the performance demands, but at the lowest price. Start being really nice to your instructor because that's the person you must please.

PART 2

A+ GUIDE TO
PC OPERATING SYSTEMS

INTRODUCTION

The labs included in Part 2 have been specifically designed to provide the student with some hands-on experience in the installation, configuration, and troubleshooting of Microsoft operating systems. In addition, there is going to be a bit of exposure to the system hardware as well, so it wouldn't hurt to have a basic understanding in that area of expertise.

In order to accomplish the exercises in these labs the classrooms will have to be equipped with some basic necessities. These include a PC for each of the students with the following minimum specifications:

- 700MHz CPU
- 128MB RAM
- 2GB hard disk drive
- Floppy diskette drive
- CD-ROM drive

- Keyboard
- Mouse
- SVGA monitor
- One available PCI slot

A network interface card will also be required. However, it is best that this not be integrated onto the motherboard. One of the exercises deals with installing that device, so the card should not already be installed into the system.

Other materials needed to complete these labs include:

- One copy each of the following operating systems:
    - MS-DOS version 5 or later (optional)
    - Windows 98 or Windows 98SE
    - Windows 2000
    - Windows XP
- A box of floppy disks
- Tapes for the drive
- A printer (or even better, one printer for each computer)
- Device drivers for the printer
- A tape drive (optional)

- A patch cord long enough to reach from the computer to the wall jack or hub

- Writing materials

- A good sense of humor

The classroom should be set up in such a way that each student will have sufficient space to work. In one lab this will involve installing a network interface card and another will involve building a peer to peer network. For those labs, a little extra workspace is in order as well.

These labs are going to provide students an opportunity to get their hands a bit dirty as they encounter a large number of the CompTIA objectives for the Core Exam. Unfortunately, since the exam is theoretical as well as practical, some objectives won't be covered. For those, you'll need to rely on your textbook. Objectives that are covered include the following:

1.1 Identify the major desktop components and interfaces and their functions. Differentiate the characteristics of Windows 9x/Me, Windows NT 4.0 Workstation, Windows 2000 Professional, and Windows XP.

1.2 Identify the names, locations, purposes, and contents of major system files.

1.3 Demonstrate the ability to use command-line functions and utilities to manage the operating system, including the proper syntax and switches.

1.4 Identify basic concepts and procedures for creating, viewing, and managing disks, directories, and files. This includes procedures for changing file attributes and the ramifications of those changes (for example, security issues).

1.5 Identify the major operating system utilities, their purpose, location, and available switches.

2.1 Identify the procedures for installing Windows 9x/Me, Windows NT 4.0 Workstation, Windows 2000 Professional, and Windows XP, and bringing the operating system to a basic operational level.

2.3 Identify the basic system boot sequences and boot methods, including the steps to create an emergency boot disk with utilities installed for Windows 9x/Me, Windows NT 4.0 Workstation, Windows 2000 Professional, and Windows XP.

2.4 Identify procedures for installing/adding a device, including loading, adding, and configuring device drivers, and required software.

3.1 Recognize and interpret the meaning of common error codes and startup messages from the boot sequence and identify steps to correct the problems.

3.2 Recognize when to use common diagnostic utilities and tools. Given a diagnostic scenario involving one of these utilities or tools, select the appropriate steps needed to resolve the problem.

3.3 Recognize common operational and usability problems and determine how to resolve them.

4.1 Identify the networking capabilities of Windows. Given configuration parameters, configure the operating system to connect to a network.

4.2 Identify the basic Internet protocols and terminologies. Identify procedures for establishing Internet connectivity. In a given scenario, configure the operating system to connect to and use Internet resources.

You will notice as you go through the labs that the same objectives will appear in different labs, again and again. This shouldn't come as much of a surprise since each of these objectives includes a number of subcomponents. So don't get tied up in the details to the point that you don't have fun and *learn!*

MS-DOS, CONFIG.SYS, AND AUTOEXEC.BAT

For many of you, this is not likely to be a hands-on lab. It's been a few years since MS-DOS has been available for purchase. If your school has copies available, then I highly recommend that the installation portion of this lab be completed as described. A hands-on installation will demonstrate just how far OS technology has come in two decades. It will also provide a more solid foundation for understanding the concepts and importance of CONFIG.SYS and AUTOEXEC.BAT.

In the event that this is not possible, skip to the sections that discuss various commands issued from the command prompt. Regardless of whether you ever see another system running DOS, you will still need a solid understanding of the command prompt in order to be a successful computer technician. *NOTE: This lab cannot be successfully completed using the command prompt of Windows XP without errors.*

The only materials you'll need for this lab are your lab computer systems and a copy of any version of MS-DOS later than 5.0.

The CompTIA objectives covered in this lab include the following:

1.1 Identify the *major* desktop components and interfaces and their functions. Differentiate the characteristics of Windows 9x/Me, Windows NT 4.0 Workstation, Windows 2000 Professional, and Windows XP.

1.2 Identify the names, locations, purposes, and contents of major system files.

1.3 Demonstrate the ability to use command-line functions and utilities to manage the operating system, including the proper syntax and switches.

EXERCISE 1: INSTALLING MS-DOS (OPTIONAL)

As much as I hate to do this, I have no choice but to make this an optional lab. If the software is available for installation, then by all means proceed. Unfortunately, this won't be an option for many. Therefore, the best you'll be able to do is follow along with the screen shots and read the instructions.

MS-DOS did not ship on CDs as do the OSs to which you've grown accustomed today. Depending on the version, DOS shipped on three to four high-density 3.5" floppy diskettes. DOS 6.22 was a short-lived version that shipped on four floppies. All other versions shipped on three. The first diskette is bootable. To install the OS, you place this first installation diskette into the floppy drive and start the machine. This diskette is configured with a default AUTOEXEC.BAT and CONFIG.SYS that first do a system check, as you see in **Figure L1.1** and then boot the computer to the initial installation screen shown in **Figure L1.2**.

From this screen, the user has several options. Pressing <Enter> causes the installation process to continue. Pressing F1 brings up a very limited help menu. F3 allows the boot process to continue—a skeleton OS is loaded to RAM, and the user can boot the machine to an A: prompt. This is useful as a troubleshooting tool in a few instances. Pressing F5 converts the display to monochrome, and F7 allows the user to set up an installation of MS-DOS that runs exclusively from floppy disks. (Yes, as amazing as it may seem, there was a time when computers did not ship with hard drives.)

Figure L1.1   First DOS does a system check to see how much horsepower you're running.

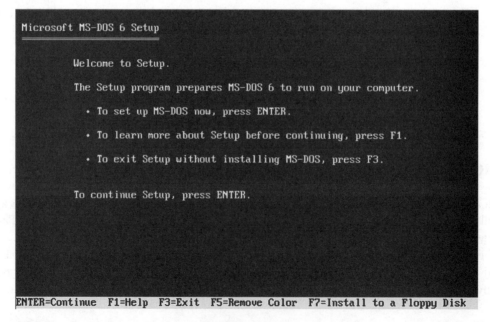

Figure L1.2   The initial installation screen of MS-DOS

Obviously, since this lab concerns the installation of DOS, you'll press <Enter>. If the hard disk onto which you are installing DOS has not been partitioned and formatted, as in the case of a brand new system or hard disk, the next screen to appear is the one shown in **Figure L1.3**. Select the option Configure unallocated disk space (recommended) and press <Enter>. By the way, if you are actually performing this installation, by now you've noticed that your mouse doesn't work. You have to make all of these selections using the arrows on your keyboard. How quaint!

This step doesn't take long at all. What is happening is that, in the background of the splash screen, the DOS installation program is using the FDISK utility to create an active partition and is writing the master boot record (MBR) to the hard drive. Next the screen shown in **Figure L1.4** tells you that your

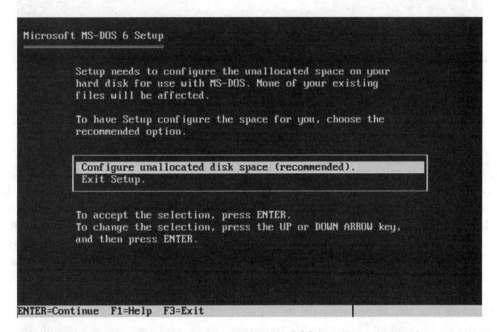

Figure L1.3   Before a new hard disk can accept a DOS installation, it must be configured.

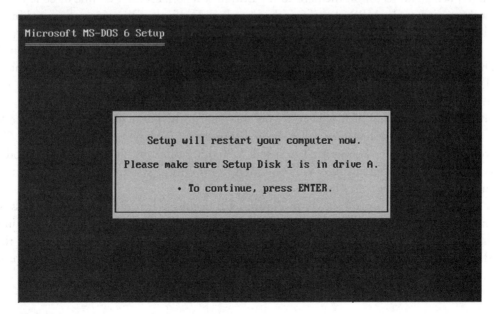

Figure L1.4   The first reboot

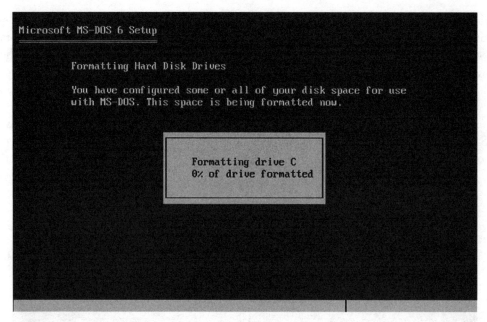

Figure L1.5 Formatting the hard disk

machine is about to reboot and that you need to leave Setup Disk One in the drive. Without that diskette in the floppy disk drive, the machine will not yet boot. There's still a lot more work left to do.

After the reboot is completed, the next step for the installation program is to format your hard disk (**Figure L1.5**). Something to keep in the back of your head throughout this lab is that MS-DOS knew no file system more complex than FAT16. (See Chapter One: An Introduction to Operating Systems in your text for a detailed discussion of file systems.) As such, the largest partition it can create is 2GB. Therefore, even if you're installing DOS onto an 80GB hard drive, the format procedure at this stage is going to give you a maximum of 2GB.

There is one final step in preparing the drive that must be done before this process begins. This step is done in the background and is completely transparent to the end user, but as a technician, you need to be aware of what is going on. During the formatting procedure, the file allocation tables (FAT) are being written to the hard disk. After the FAT is completed, the DOS installation program knows exactly what cylinder, head, and sector on the hard disk partition will contain the first line of code for the OS. It writes a short entry to the MBR identifying that location. This is the OS pointer discussed in Chapter One of the text.

Next you have the option of changing the default configuration for the DOS installation (**Figure L1.6**). There isn't a whole lot involved in this process. You can set the date and time, the country where you reside, and your choice of keyboard layout. Any other configuration changes have to wait until the installation is completed.

Before the actual copying of files begins, you are going to be given the opportunity to select the directory where the DOS system files will be stored (**Figure L1.7**). The default directory is C:\DOS. Unless you have some overpowering need to install to another directory, it is generally a good idea to select the default. Some programs look only to C:\DOS for critical system files and don't run if they aren't found in the first place they look.

After that the file copy process shown in **Figure L1.8** begins. When the File Copy process is about 20 percent complete, you will be prompted to insert Setup Disk Two by a screen like the one in **Figure L1.9**. This will happen yet again when the copying is about 56 percent done. Just insert the disks requested. If you accidentally insert the wrong diskette, Installation will tell you. The floppies are digitally labeled, and Setup reads the label to make sure it has the correct file source.

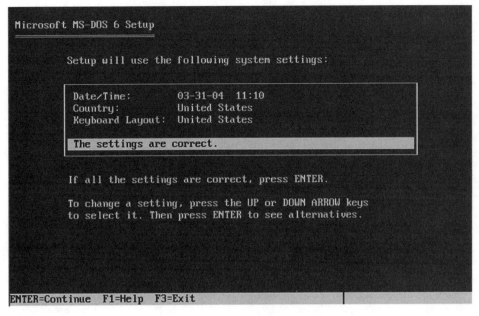

Figure L1.6 Configuring DOS

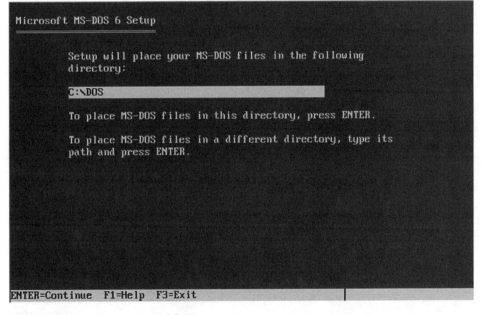

Figure L1.7 Selecting the DOS directory

When the last file has been copied to your hard drive, you've reached the final stage of the DOS installation. You must reboot the system one more time. You want it to boot to the hard disk, so, as the Setup program tells you in **Figure L1.10**, you need to remove the last floppy diskette from the drive and press the <Enter> key to continue.

MS-DOS is now installed on your computer. It will now boot to the lovely command prompt each and every time your start your computer.

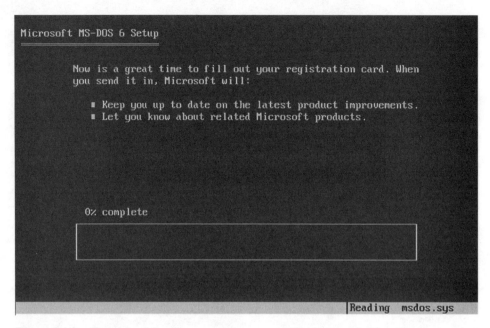

Figure L1.8 Throughout the file copy process, you'll be bombarded by advertisements for a product you already bought.

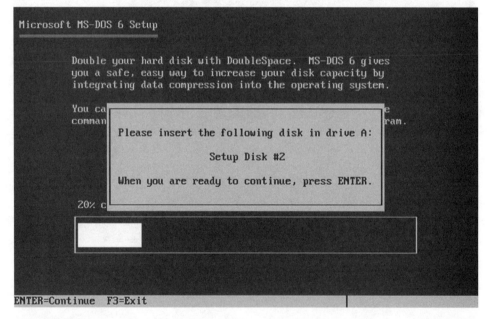

Figure L1.9 When the File Copy process is about 20 percent complete, and once again when it is about 56 percent complete, you will be prompted to change the diskette in the drive.

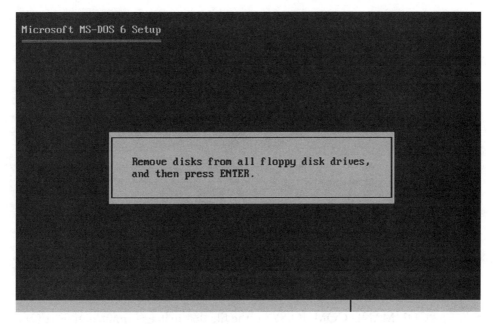

Figure L1.10   The final reboot

EXERCISE 1 REVIEW

1. How many diskettes did MS-DOS 6.22 ship on? How about the other versions?

2. How many times does the computer reboot during the installation process?

EXERCISE 2: CONFIG.SYS AND AUTOEXEC.BAT

Two files that were almost always necessary to the MS-DOS configuration were CONFIG.SYS and AUTOEXEC.BAT. CONFIG.SYS, as its name implies, is a file that defines the system configuration. AUTOEXEC.BAT defined what programs would launch automatically each time the machine started, and then remain RAM-resident the entire time, even if for any reason the program was shut down. Programs of this nature were called Terminate and Stay Resident (TSR) applications. Both of these files could be used to fine-tune the system configuration.

If you recall from Exercise 1, I mentioned that the configuration allowed during installation was rather minimal. CONFIG.SYS and AUTOEXEC.BAT allow third-party device drivers, configuration parameters, and TSRs to be selected and tweaked by the user. Both of these files load in advance of any user applications, although AUTOEXEC.BAT can be used to automatically order user applications to be loaded.

If you didn't have the luxury of installing MS-DOS onto computer systems in Exercise 1, the following exercises can be done from any Microsoft-based computer using the command prompt. On a Windows 98 computer, you can either click Start→Programs→MS-DOS Prompt or you can click Start→Run and type **command** in the run line. On Win2K or WINXP machines, select Start→Run and type **CMD** into the field.

EXERCISE 2A: CONFIG.SYS

If you recall from Chapter Two: MS-DOS and the Command Prompt from your text book, DOS has a strict order in which it loads. Three files constitute the core of the OS. These files are IO.SYS,

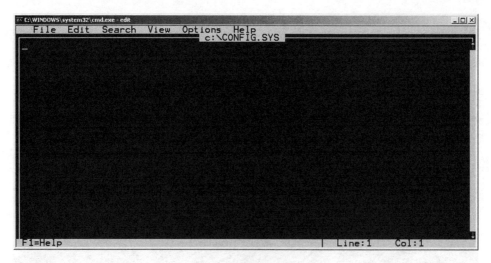

Figure L1.11 The XP incarnation of the old DOS Editor

MSDOS.SYS, and COMMAND.COM. IO.SYS is the file that manages input/output (I/O) operations. You probably already figured that out from its name. A key task of I/O for any computer system is talking to peripherals installed on the system. This includes devices that were installed after the fact, such as sound cards, modems, SCSI adapters, and a host of other devices that didn't come equipped on the original computer. During the boot process, before anything else can happen, I/O control must be established. Therefore, IO.SYS is the first file to load. It locates and loads CONFIG.SYS. CONFIG.SYS is where device drivers, both real and virtual, are loaded, as well as a number of configuration parameters. In this section I'll cover some typical commands to be loaded into CONFIG.SYS.

In the old days of DOS these files were edited using the MS-DOS Editor. This was launched by typing **Edit** at the command prompt. Oddly enough, that little utility has never gone away. **Figure L1.11** shows the Edit utility for WINXP. For editing CONFIG.SYS and AUTOEXEC.BAT, either the DOS editor or the Notepad utility in Windows should always be used. They do not add extraneous formatting characters, as do Wordpad or MS Word.

For the rest of this exercise and throughout the next, I'll be having you use the command prompt editor for creating these files. If you are performing these exercises on working machines, you might want to save any pre-existing CONFIG.SYS files on the system. From the command prompt, type **CD** to get to the root directory. Next, type **REN CONFIG.SYS CONFIG.BAK** in order to rename the file. Type **REN AUTOEXEC.BAT AUTOEXEC.BAK** to rename AUTOEXEC.BAT.

To start, simply type the first line I provide into the Editor and then save the file as C:\CONFIG.SYS. When you're finished, make sure you save the final version. Now let's look at some of the commands typically found in CONFIG.SYS. In order to emphasize where spaces belong and where they don't in the next few pages, I will be placing three spaces in place of a single space. This is not recommended in real practice.

FILES= This statement specifies the maximum number of individual file handles DOS will generate. Since DOS can only open a file if there is a file handle available, this dictates how many files can be open. With no entry, the default setting is 8. The maximum number allowable is 255. Frequently the default number of files DOS could open was far too small. Some applications adjusted this value for the end user during installation, increasing it to the necessary level. If not, a minimum value of 32 was recommended. It was important not to get carried away. Increasing the number in the FILES= statement increased the amount of conventional memory used. **Table L1.1** lists the amount of memory used for some common values.

The proper syntax for the FILES= statement is

`FILES=32`

Type that into CONFIG.SYS and save your file.

BUFFERS= DOS needs an area of reserved memory it can use for storing data that is in transit. I/O operations rarely occur in a single clock cycle. Therefore, a stream of data can only flow as fast as the receiving device can take it. This is true of commands as well as data. If a system is configured to use high memory, 512 bytes of conventional memory will be allocated as a "window" for buffered data as it moves from conventional to high memory. If high memory is not used, each buffer will allocate 532 bytes of conventional memory. By default, the number of buffers allocated by DOS varies with system configuration. **Table L1.2** shows the default allocation of buffers by MS-DOS.

The proper syntax for loading buffers is

`BUFFERS=20`

Type that line into CONFIG.SYS and save the file.

STACKS= Hardware interrupts require a certain amount of conventional memory for their interrupt drivers. The STACKS= statement is going to contain two values, separated by a comma. The first value dictates the number of stacks, and the second determines the size of each stack in bytes. A valid range for the number of stacks is from 8 to 64. Stacks can be a maximum of 512 bytes. If MS-DOS detects a CPU more recent than a 80286, it assigns a default value of 9,128, or nine stacks of 128 bytes. Most systems are more stable with nine stacks of 256 bytes. The correct syntax for the line is

`STACKS=9,256`

Table L1.1 Conventional Memory Used by the FILES= Statement

Files Value	Bytes Consumed
8 (Default)	192
10	496
15	608
20	896
25	1200
30	1488
35	1776
40	2080
45	2368
50	2672
55	2960
60	3260
65	3552
70	3856
75	4144

Increasing the number of files open in CONFIG.SYS consumed additional conventional memory.

Table L1.2 Default Buffers and Memory Usage in MS-DOS

Configuration	Default Buffers	Bytes Used
<128K of RAM, 360K disk	2	1064
<128K of RAM, 720K disk (or larger)	3	1596
128K to 255K of RAM	5	2672
256K to 511K of RAM	10	5328
512K to 640K of RAM	15	7984

When buffers load, increasing the number depletes the supply of available conventional memory.

Add this line to CONFIG.SYS and save the file.

DEVICE= and **DEVICEHIGH=** These are the commands that load third-party device drivers during the boot process of MS-DOS. DEVICE= loads the drivers into conventional memory, while DEVICEHIGH= loads them into high memory. For the latter to work, the first device to be loaded has to be a DOS virtual device called HIMEM.SYS. In order to make use of extended memory, this line has to be followed by another that loads a program called EMM386.EXE. To properly load, the entire path to the file has to be specified in the line. After the name of the device driver, additional parameters specific to the drive can be loaded. This might include bits of information such as what IRQ the device is configured for, a specific I/O address, or a DMA channel. Parameters are preceded by a forward slash. An example used by HIMEM.SYS is the parameter TESTMEM. The options are ON or OFF. Since the vast majority of people using this lab manual will be playing with Windows machines and not DOS, this example is based on the Windows OS. If you were fortunate enough to be able to install DOS, replace \WINDOWS\SYSTEM32 with \DOS. The proper syntax is

```
DEVICE=C:\WINDOWS\SYSTEM32\HIMEM.SYS   /TESTMEM:OFF
```

Add that line to CONFIG.SYS and save your file. Remember that where I used three spaces, you only need one. Since you need EMM386.EXE to be running before you can use expanded or extended memory, also add the line:

```
DEVICE=C:\WINDOWS\EMM386.EXE
```

CAUTION: Do not try this in XP. Add this to CONFIG.SYS and save the file.

LASTDRIVE= By default, DOS looks at the letter of the last hard disk (either physical or logical) installed and adds one more drive, up to drive letter E. By adding a LASTDRIVE statement, drive letters up to Drive Z are possible. The correct syntax is:

```
LASTDRIVE=Z
```

Add this line to CONFIG.SYS and save your file.

NUMLOCK= People who use the number pad on their keyboard for numbers always want this setting to be ON. Gamers who use those keys as arrows for their characters and triggers for their weapons want it OFF. This line in CONFIG.SYS overrides any setting in BIOS. The correct syntax is:

```
NUMLOCK=OFF
```

Or ON, whichever, you prefer. You decide and add the line to your CONFIG.SYS. Save the file.

There are other commands that can be added to CONFIG.SYS, but these are the most common. For a more detailed discussion, refer to Chapter Two in the text book.

Exercise 2a Review

1. You've just installed a new DOS program and the first time you run it, you get the message that says "Insufficient File Handles. Application Terminated." What line in CONFIG.SYS would you change to fix this problem?

2. You're running MS-DOS and are frequently getting a message that says "STACK OVERFLOW," and the system shuts down. What line can you change to fix this problem?

EXERCISE 2B: AUTOEXEC.BAT

AUTOEXEC.BAT is kind of a catch-all file that technicians can use to add any number of different things. Device drivers that run as executables can be loaded from here, as can any program or batch file that your little heart desires. There are also some system parameters that load as commands. Here is a look at a few AUTOEXEC.BAT functions.

As with Exercise 2a, open your command prompt editor and create a new file called C:\AUTOEXEC.BAT. As before, if this is a functioning system that you want to put back the way it was, rename the existing AUTOEXEC.BAT to AUTOEXEC.BAK.

@ECHO OFF Many technicians and end users well-versed in MS-DOS frequently place this as the first line in an AUTOEXEC.BAT file. This line prevents commands in AUTOEXEC.BAT from flashing up on the display as they are read and executed. Proper syntax is

```
@ECHO OFF
```

Add this line to your AUTOEXEC.BAT and save the file.

PROMPT This command dictates what the command prompt will look like. There is no equal sign used after this command as in some of the previous examples. There are a number of parameters that define different things, and each of these parameters must be preceded by a dollar sign ($). **Table L1.3** is a list of available PROMPT parameters and what they do.

The proper syntax for a PROMPT command is

```
PROMPT  $P$G$T$D
```

As before, where I've got three spaces for emphasis, you need but one. Add this line to your AUTOEXEC.BAT and save your file.

Table L1.3 Prompt Parameters and Their Results

Parameter	Result
$T	Current time
$D	Current date
$P	Current drive and path
$V	Windows version number
$N	Current drive
$G	> (greater-than sign)
$L	< (less-than sign)
$B	| (pipe)
$H	Backspace (erases previous character)
$E	Escape code (ASCII code 27)
$_	Carriage return and linefeed

In the days of DOS, one of the ways you advertised yourself as a true geek was to make a customized command prompt.

PATH= In the days of DOS, everything was done from a command prompt. Typing a command at the prompt either worked, or it didn't. One of the things that dictated what happened was your path statement. If your prompt looks like C:\DOS\COMMAND, that indicates that you have browsed to the COMMAND subdirectory of the DOS directory. If you type a command at that point and have no PATH statement in AUTOEXEC.BAT, DOS will look in the C:\DOS\COMMAND directory for that command. If the command does not reside in that directory, the user gets the infamous "BAD COMMAND OR FILE NAME" error message. If a PATH statement exists in AUTOEXEC.BAT, after looking in C:\DOS\COMMAND, COMMAND.COM will next search each directory listed in the PATH statement. Most applications are programmed to automatically add themselves to the PATH statement when they install. Each directory in the statement is separated by a semicolon. The correct syntax is

```
PATH=C:\;\DOS;\WINDOWS
```

Add this line to AUTOEXEC.BAT and save your file.

LH This is an acronym for Load High. LH followed by any executable command loads that executable into upper memory, provided there is sufficient space in upper memory to do so and providing that HIMEM.SYS and EMM386.SYS are properly loaded. An example of this command in use, with proper syntax is

```
LH C:\DOS\MOUSE.COM
```

Since it is highly unlikely that you will actually have a DOS directory on the machine you're using, I will *not* have you copy this file to your AUTOEXEC.BAT.

Now that you have both a newly configured CONFIG.SYS and AUTOEXEC.BAT files loaded, reboot your machine and watch for changes. The most noticeable will be your command prompt when you open a command prompt window.

Exercise 2b Review

1. What was the purpose of @ECHO OFF?
2. What would be the effect of adding the line PROMPT PG to your AUTOEXEC.BAT?

Exercise 3: Troubleshooting CONFIG.SYS (optional)

One of the things about CONFIG.SYS is that, like any other computer code in your system, the system reads the lines in this file quite literally. Any misspelling of a command or device driver will result in errors, and if the path to a particular file isn't correct, you'll get errors. Also, the lines are loaded in the order they appear in the file. If you recall from Exercise 2a, I mentioned that in order to use high memory you have to have a file called HIMEM.SYS loaded. In order for you to use extended memory, EMM386.EXE has to be loaded. They also have to be loaded in the correct order. If you were to reverse lines four and five of the CONFIG.SYS file you created earlier in this lab, the next time you booted your machine you would get an error message during boot up that reads "Error in config.sys, line 4." It won't tell you what the error is, but since it tells you what line the error is in, you should have no difficulty looking at the specific line for errors. Problems to look for include proper spelling, proper syntax, proper placement in the file, and proper use of parameters.

EMM386.EXE requires the services of HIMEM.SYS to properly function. If HIMEM.SYS isn't already loaded when EMM386.EXE attempts to load, it will fail and deliver a message.

Adding any line relating to EMM386.EXE in Windows XP will result in errors. There is no expanded memory support in XP, and this file no longer exists. Let's look at how it works.

1. Open your CONFIG.SYS file in the DOS Editor.

2. Edit line four to read DEVICE=HIMEM.SYS.

3. Restart your machine.

You'll immediately be treated to the Error in CONFIG.SYS, Line X error message. HIMEM.SYS is not located in the root directory and, therefore, could not be located without the full path.

EXERCISE 3 REVIEW

1. If you are running a Windows machine and you have an error in CONFIG.SYS, how will that error affect your machine? (Note: You'll have to refer to the textbook, Chapter Two, to answer this question.)

2. If the PATH statement in AUTOEXEC.BAT contains the correct path to HIMEM.SYS, will CONFIG.SYS then be able to locate the file even if you haven't typed the full path into the CONFIG.SYS line?

LAB REVIEW

1. You have a device driver located in the C:\TOY directory called TOY.SYS. How would you write a line to load that driver at startup?

2. In what file should you place the above-mentioned line?

3. What utility is used to prepare a hard disk for Microsoft OSs using any version of FAT?

4. You are playing an old DOS game and constantly get STACK OVERFLOW errors the instant the machine crashes. What line might fix this problem and where would you put it?

LAB SUMMARY

For the average Windows user, this entire lab appears on the surface to be about as useful as an ejection seat in a helicopter. Nobody runs DOS anymore, and Windows makes no use of CONFIG.SYS or AUTOEXEC.BAT. So why waste your time?

As of this writing, there are still a number of companies running proprietary DOS applications that for one reason or the other, they can't port to Windows. For those apps to work, a viable CONFIG.SYS and/or AUTOEXEC.BAT file must be present and properly configured. What a wonderful impression you're going to make on your new boss if you have to look him or her right in the eye and say, "I haven't got a clue how to do that."

Preparing a Hard Disk the Old-Fashioned Way

Starting with this lab, you're going to begin looking at operating systems that still have a substantial base of installations around the world. An unofficial estimate at the time of this writing is that there are still in excess of 20 million computers out there running some version of Win9x. Although there is a possibility that you may never be called on to install the product, there is an even stronger possibility that you *will* be called on to service it. You might as well get the hang of installing it as long as you're going to all this effort anyway.

The only materials you'll need for this lab are your lab computers, Win98 CD and Startup diskette, and any third-party drivers required for proprietary hardware.

The CompTIA objectives covered in this lab include the following:

1.3 Demonstrate the ability to use command-line functions and utilities to manage the operating system, including the proper syntax and switches.

1.4 Identify basic concepts and procedures for creating, viewing, and managing disks, directories, and files. This includes procedures for changing file attributes and the ramifications of those changes (for example, security issues).

1.5 Identify the major operating system utilities, their purpose, location, and available switches.

Preparing the Hard Disk

In the next hour to an hour and a half, you'll be going through an actual installation of Windows 98. I'll take you through the process step by step and explain what's happening along the way. I'll also show you a couple of things you might miss if all you ever do is accept defaults. However, in most cases, before the OS can be installed, the computer's hard disk has to be partitioned and formatted. Exercise 1 covers this aspect of the installation.

EXERCISE 1: THE FDISK UTILITY

For this exercise, each student needs to have access to a lab machine and a boot diskette with CD-ROM support. If you used the Windows 98 Startup Disk utility or downloaded the boot image from www.mwgraves.com, then you have what you need. So let's install Windows. (*Note:* Windows 98 ships on a bootable CD, so if for any reason, a boot diskette is not available, Setup can be started from the CD as long as the machine's CMOS is configured to boot to CD-ROM. The purpose of booting to the floppy in this exercise is to point out a minor difference in the process.)

Insert the boot diskette into the floppy diskette drive and start the machine. If this is the first time you've ever booted to floppy, don't be concerned about the amount of time it takes. That's normal. Using a boot diskette created from Win98, you'll get a Startup menu similar to the one in **Figure L2.1**.

Since you want to configure your hard drive before you do anything else, there's no point in waiting for all those CD-ROM drivers to load. After it boots to the A:\ prompt, perform the following steps:

1. At the A: prompt, type **fdisk** and then press the <Enter> key. The opening screen will appear (**Figure L2.2**). All that fancy explanation is simply asking you whether you want to format your drive to FAT16 or FAT32. If you choose Y (yes) when it asks whether you want to use large disk support, you have selected FAT32. If you press N (no), you will select FAT16. If for any reason you need FAT16, you should press N. For the purposes of this (and later) exercises, press Y.

2. The FDISK Options screen appears next (**Figure L2.3**). It is from this screen that the various tasks are selected. Which task you select depends entirely upon what you're trying to do. On a brand new unprepared hard drive, you would choose Option 1: Create DOS partition or Logical DOS Drive. If you are working with a disk that has previously contained data, you would want to start by selecting Option 4: Display partition information. Option 5: Change current fixed disk drive, will appear only in systems that have more than one physical hard disk installed. FDISK can recognize when there is only a single hard disk and, in that case, Option 5 will not appear. For this exercise, select Option 4.

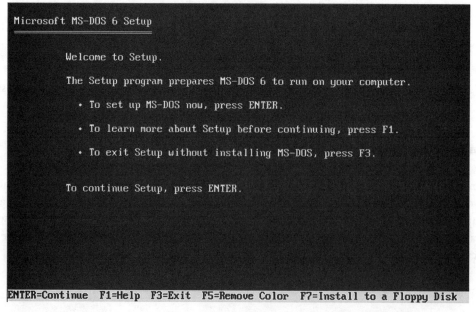

```
Microsoft MS-DOS 6 Setup
────────────────────────

     Welcome to Setup.

     The Setup program prepares MS-DOS 6 to run on your computer.

          • To set up MS-DOS now, press ENTER.

          • To learn more about Setup before continuing, press F1.

          • To exit Setup without installing MS-DOS, press F3.


     To continue Setup, press ENTER.




 ENTER=Continue  F1=Help  F3=Exit  F5=Remove Color  F7=Install to a Floppy Disk
```

Figure L2.1 Booting a machine to a Startup diskette created by Win98 offers a boot menu with three options.

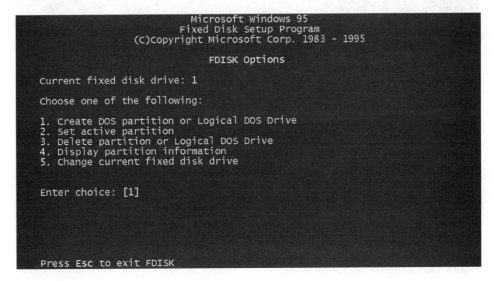

```
Your computer has a disk larger than 512 MB. This version of Windows
includes improved support for large disks, resulting in more efficient
use of disk space on large drives, and allowing disks over 2 GB to be
formatted as a single drive.

IMPORTANT: If you enable large disk support and create any new drives on this
disk, you will not be able to access the new drive(s) using other operating
systems, including some versions of Windows 95 and Windows NT, as well as
earlier versions of Windows and MS-DOS. In addition, disk utilites that
were not designed explicitly for the FAT32 file system will not be able
to work with this disk. If you need to access this disk with other operating
systems or older disk utilities, do not enable large drive support.

Do you wish to enable large disk support (Y/N)...........? [N]
```

Figure L2.2   The opening screen of FDISK

```
                   Microsoft Windows 95
                   Fixed Disk Setup Program
              (C)Copyright Microsoft Corp. 1983 - 1995

                       FDISK Options

Current fixed disk drive: 1

Choose one of the following:

1. Create DOS partition or Logical DOS Drive
2. Set active partition
3. Delete partition or Logical DOS Drive
4. Display partition information
5. Change current fixed disk drive


Enter choice: [1]




Press Esc to exit FDISK
```

Figure L2.3   The FDISK Options screen

3.   The screen shown in **Figure L2.4** will appear. The information contained in that screen may vary from class to class and from computer to computer, depending on what was previously installed on that hard drive. If it's a brand new hard drive, there should be a message telling you that no DOS partitions exist.

4.   Press the <Esc> key to return to the FDISK Options screen. If your systems contained previous partitions, you need to delete them. To do so, press Option 3: Delete partition or Logical DOS Drive. This will bring up a screen like the one shown in **Figure L2.5**.

5.   If there are extended partitions on your drives, you must first select Option 3: Delete Logical DOS Drive(s) in the Extended DOS Partition and delete each logical drive on the partition. Next you need to delete the extended partition, and finally you can proceed to Option 1: Delete Primary DOS Partition.

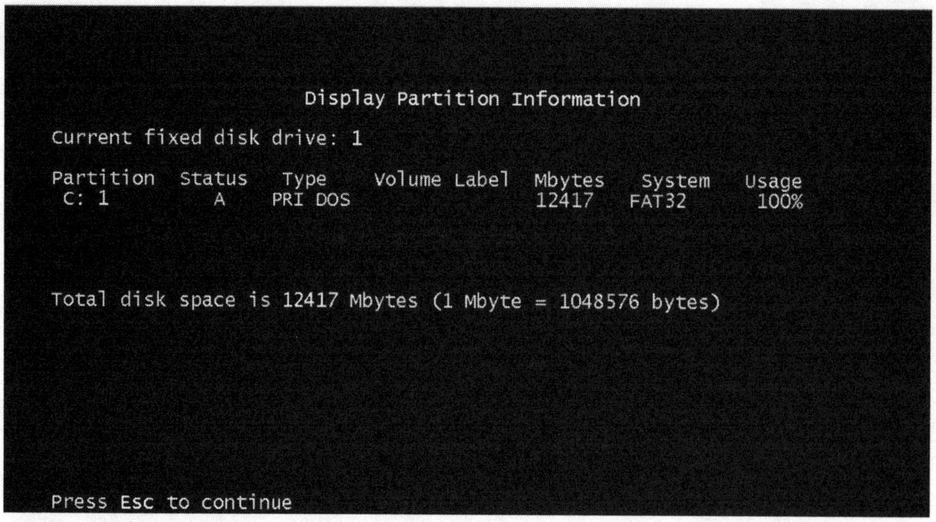

```
                    Display Partition Information
Current fixed disk drive: 1

Partition  Status   Type    Volume Label  Mbytes   System   Usage
  C: 1       A     PRI DOS                 12417    FAT32    100%




Total disk space is 12417 Mbytes (1 Mbyte = 1048576 bytes)






Press Esc to continue
```

Figure L2.4   Viewing partitions in FDISK

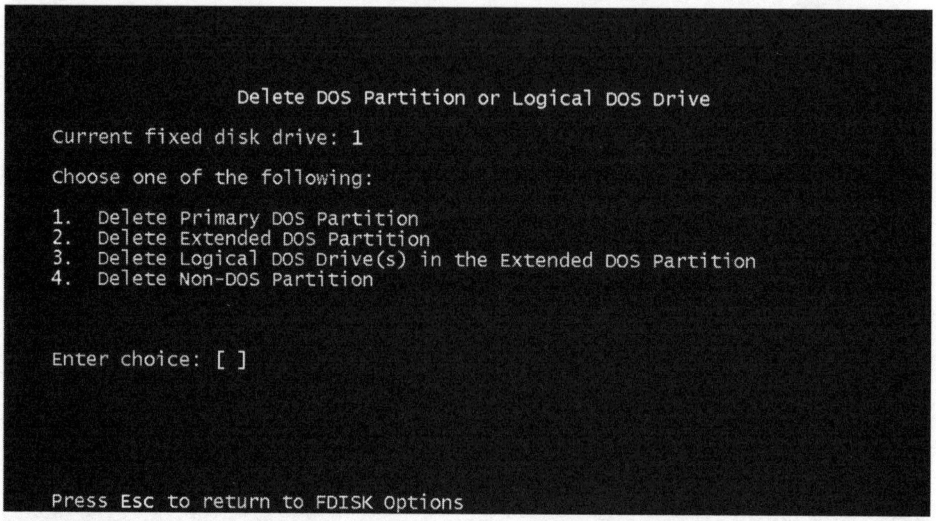

```
                Delete DOS Partition or Logical DOS Drive
Current fixed disk drive: 1

Choose one of the following:

1.  Delete Primary DOS Partition
2.  Delete Extended DOS Partition
3.  Delete Logical DOS Drive(s) in the Extended DOS Partition
4.  Delete Non-DOS Partition


Enter choice: [ ]




Press Esc to return to FDISK Options
```

Figure L2.5   Deleting a partition in FDISK.

6. This brings up the screen shown in **Figure L2.6**. You are warned that to continue will delete all data on your drive. Select Y and a new field will open up prompting you to enter the Volume Name. If your drive has a volume name, you must enter it exactly as it appears on the screen. You will then be given yet another option to bail out when it asks whether you're really sure. Press Y to delete the partition, and now the previous data on the hard drive is history. This is when far too many technicians scream "**No, wait! That was the wrong drive!!!**"

7. Press <Esc> to return to the FDISK Options menu. Now you want to select Option 1: Create DOS partition or Logical DOS Drive. This will bring up the screen shown in **Figure L2.7**. Before you can create any extended partitions, a primary DOS partition must first be created. Press Option 1: Create Primary DOS Partition. You will be asked whether you want to use all available space and to make the partition active. Select Y.

8. FDISK will then quickly scan the disk for obvious flaws, calculate available space, and create the primary partition. Press <Esc> twice to exit FDISK. You will be prompted that for the changes to take effect, you must restart your machine. Make sure the boot diskette is still in Drive A:, select Y, and let the machine reboot. You are now ready for Exercise 2.

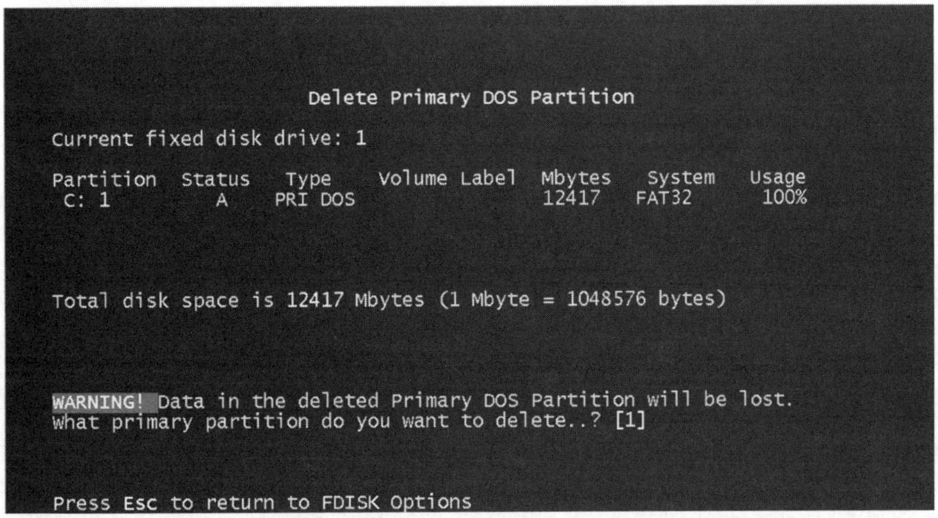

```
                        Delete Primary DOS Partition

Current fixed disk drive: 1

Partition   Status    Type    Volume Label   Mbytes    System    Usage
   C: 1       A      PRI DOS                  12417     FAT32     100%



Total disk space is 12417 Mbytes (1 Mbyte = 1048576 bytes)



WARNING! Data in the deleted Primary DOS Partition will be lost.
What primary partition do you want to delete..? [1]



Press Esc to return to FDISK Options
```

Figure L2.6   You get one last chance to bail out!

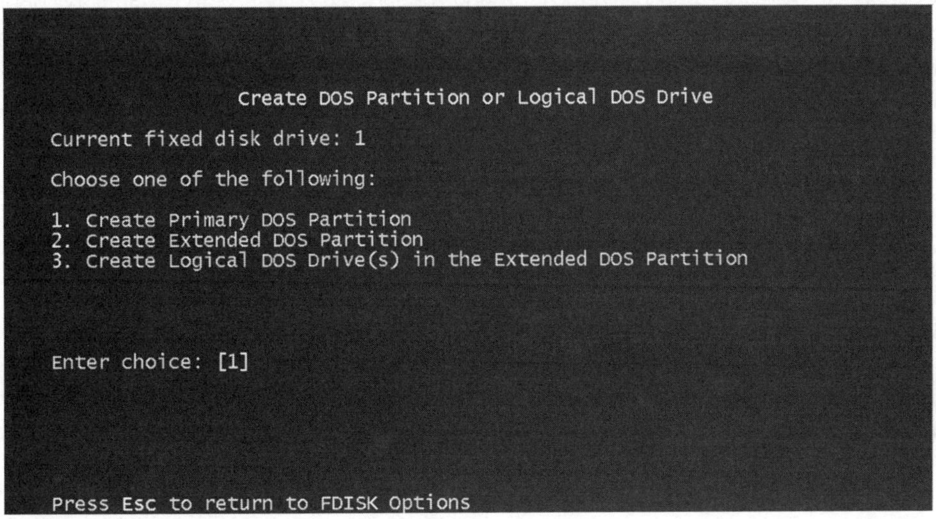

```
                   Create DOS Partition or Logical DOS Drive

Current fixed disk drive: 1

Choose one of the following:

1. Create Primary DOS Partition
2. Create Extended DOS Partition
3. Create Logical DOS Drive(s) in the Extended DOS Partition



Enter choice: [1]





Press Esc to return to FDISK Options
```

Figure L2.7   Creating a DOS partition

EXERCISE 1 REVIEW

1. Why is it a good idea to view partitions on a previously existing hard disk before making any permanent changes?

2. If you divided your hard disk into multiple partitions using FDISK, what additional step must you complete before you will have a bootable drive?

3. Using FDISK, you want to remove an extended partition, but it won't let you. What does the error message tell you that you must do before removing extended partitions?

4. You created two separate partitions on the same hard disk, using the % Disk Space method of allocating disk space to each partition. Now what must you do before you have a bootable system?

5. In the opening screen of FDISK, you selected N when asked whether you wanted to enable Large Disk Support. Just what did you do to the system when you made that selection?

SOME COMMENTS ON FDISK

Keep in mind that FDISK is an older utility and, as such, is subject to some limitations. For example, a partition created by a non-Microsoft OS such as Linux frequently can't be recognized. When you view partitions, you see a Non-DOS Partition listed, but when you try to remove it, you are informed that no Non-DOS partitions exist. It can drive you nuts. Even Microsoft OSs occasionally do something to the MBR that prevents FDISK from being able to work with that disk. I frequently work in classroom environments where the OS of a given system changes from day to day. I discovered that in some cases, Windows 2000 Advanced Server would make the disk unrecognizable after the students had converted the drives to Dynamic Disks. This is an option in Windows 2000 that applies advanced properties to the drive that earlier OSs don't know how to deal with. Oddly enough, I never encountered that problem in classrooms using Windows 2000 Professional.

Along a different line, should you with to create multiple partitions using FDISK, you have two options for dividing your disk into partitions. In Step 8, instead of selecting Y when asked whether you want to use all available space, select N. Now you will be encouraged to enter how much of the drive you want used for the primary partition. If you simply type in a number, such as 150, you are defining how many megabytes of space that partition should contain. A number followed by the percent sign is obviously a percentage. If you type **50%**, it will take exactly half of the available space and assign it to the primary partition. Either way you choose, the result of your actions will be that your primary partition is *not* set as active by default. You'll have to go back to the initial FDISK menu, select Option 2: Set active partition, and select which of the partitions you've created should be the active one.

EXERCISE 2: FORMATTING A HARD DISK

Partitioning the hard disk created a new master boot record, but the disk still isn't usable. You now need to generate the file allocation tables. This is done when you format the hard disk. So let's do it!

1. The first thing you need to do is determine what your new drive letter is going to be. If this is a second drive in your system, you don't want to be formatting Drive C: now do you? That's when the View Partition Information option you saw in the previous exercise comes in handy. If it is the first and/or only drive in the system, then you simply type **format C:** at the command prompt.

2. As you see in **Figure L2.8**, you are warned that continuing will destroy all data on the drive. Do you wish to proceed? Well, as I was preparing this lab, I was using the same computer I'm using to write with. So I chose to press N for no. Go figure. You should press Y for yes and sit back and wait. Depending on the size of the hard drive, your age, and how long it is until lunch, this can take anywhere from several minutes to the rest of your natural life. When it finishes, you're finished with this lab.

EXERCISE 2 REVIEW

1. Why does the drive have to be formatted before you can use it?

2. Referring back to the textbook for reference, why is it possible to retrieve data from a hard disk that has just been formatted?

SOME COMMENTS ON FORMATTING

The format command used by DOS and Windows is known as a high-level format, or sometimes as the operating system format. A low-level format is done at the factory, which maps out the sectors and tracks on each platter. Some older versions of BIOS contain an option for Low Level Format. This is for SCSI or older MFM hard drives **only** and should **never** be used on an IDE drive. Using that utility on an IDE drive will turn it into a paperweight. Some of the better hardware diagnostic utilities include a low-level format for IDE drives. This might be useful in the event that a failed attempt to install an operating system has left Track 0, Sector 1 unreadable. It should, however, be used only as a last resort.

Also, on the initial format of a newly partitioned hard drive, it can take quite some time to format the drive. What the format utility is doing is going out to each file allocation unit (FAU) on the drive, testing its integrity and then going back and writing the FAT entry. Any bad FAUs will be marked as such and will not be reported to the OS. Subsequent formats of that disk can be much quicker. Using the command **format /q** will perform a quick format that only rewrites the FAT without doing the surface scan.

Something to keep in mind when reinstalling operating systems is that it is frequently a good idea to FDISK the hard disk before formatting an existing drive. The reason for this is that the operating system owns the FAT. Reinstalling the same OS can result in unpredictable problems.

```
A:\>format c:

WARNING, ALL DATA ON NON-REMOVABLE DISK
DRIVE C: WILL BE LOST!
Proceed with Format (Y/N)?
```

Figure L2.8   Formatting the hard disk.

LAB REVIEW

1. Using information provided in this lab along with information from the text, explain just what is going on during the FDISK routine when you create a primary DOS partition.
2. Using knowledge you gained in this lab as well as from the text book, explain what happened to the MBR when you divided the disk into multiple partitions.

3.   Explain what is going on during the format process and just why it takes so much longer to perform an initial format than it does when you select the `format /q` option.

4.   Using information from this lab and from the text, explain the difference between a low-level format and an operating system format.

5.   *For extra credit:* Even after FDISKing and formatting the hard drive, there is still more work that must be done by the OS during its installation. What might that be?

LAB SUMMARY

Well, now you know that either replacing or installing a new hard drive isn't simply a matter of bolting in the new drive and hooking up a couple of cables. There's a certain degree of drive preparation that needs to be done as well. It's all part of the job.

INSTALLING WINDOWS 98

Well, you're finally ready to install the OS. You have a properly FDISKed and formatted hard disk, and you're raring to go. Depending on the speed of your lab computers, the following exercise will take anywhere from 40 minutes to as much as a couple of hours.

The only materials you'll need for this lab are your lab computers, Win98 CD, and any third-party drivers required for proprietary hardware.

The CompTIA objectives covered in this lab include the following:

1.1 Identify the major desktop components and interfaces and their functions. Differentiate the characteristics of Windows 9x/Me, Windows NT 4.0 Workstation, Windows 2000 Professional, and Windows XP.

1.2 Identify the names, locations, purposes, and contents of major system files.

2.1 Identify the procedures for installing Windows 9x/Me, Windows NT 4.0 Workstation, Windows 2000 Professional, and Windows XP and bringing the operating system to a basic operational level.

EXERCISE 1: INSTALLING WINDOWS

For this exercise, each student needs to have access to a lab machine and a boot diskette with CD-ROM support. If you used the Windows 98 Startup Disk utility or downloaded the boot image from www.mwgraves.com, then you have what you need. So let's install Windows. (*Note:* Windows 98 ships on a bootable CD, so if for any reason, a boot diskette is not available, Setup can be started from the CD as long as the machine's CMOS is configured to boot to CD-ROM. The purpose of booting to the floppy in this exercise is to point out a minor difference in the process.)

Insert the boot diskette into the floppy diskette drive and start the machine. If this is the first time you've ever booted to floppy, don't be concerned about the amount of time it takes. That's normal.

The first thing you'll see is a list of the different options for starting the machine. **Figure L3.1** illustrates these options. They include starting the computer with or without CD-ROM support and viewing the help file, which is incredibly useful. Select Option 1: Start computer with CD-ROM support.

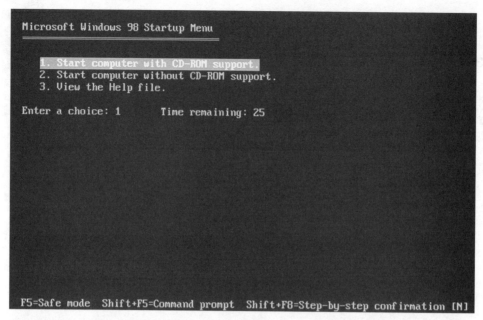

Figure L3.1 The boot menu offered by the Win98 Startup diskette

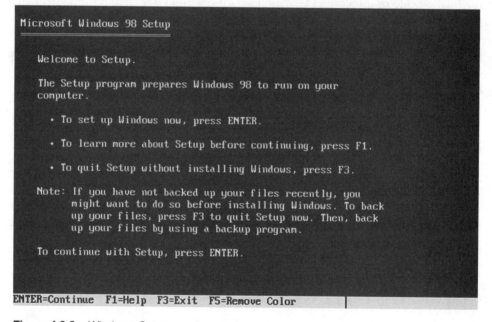

Figure L3.2 Windows Setup

When you boot a computer using the Win98 Startup diskette, a ramdisk will be created. A ramdisk is an area in physical RAM that is treated exactly as if it were a hard disk. It is divided into 512-byte sectors just like a hard disk, and a file allocation table is created for that disk. The Windows Utilities will be loaded to this disk for faster performance. This can throw off the drive letters from what you might be expecting, because the ramdisk will become drive C: until your newly FDISKed hard disk has been formatted. After the hard disk is formatted, the ramdisk will become drive D:, moving your CD-ROM up to Drive E. Therefore, in most cases, your next step is to type **E:\SETUP**. If your lab machines have multiple hard disks, this may vary.

1. This brings up the Windows 98 Setup screen (**Figure L3.2**). You have the option of starting the Windows installation, learning more about Windows setup (but not more than you'll learn

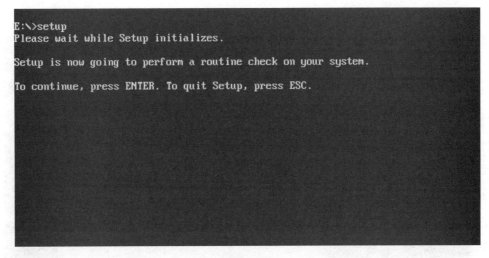

Figure L3.3   Before Setup continues, it performs a routine check on your system to make sure you have sufficient RAM and hard disk space.

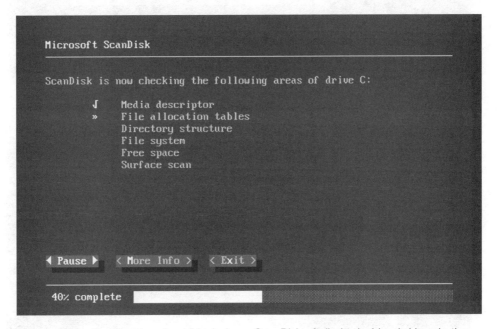

Figure L3.4   The "routine check" includes a ScanDisk of all physical hard drives in the system.

in this lab), or bailing out. The bottom of the screen tells you the press <Enter> to continue, F1 for help, F3 to quit setup, or F5 to remove color. Press <Enter>.

2. You will be notified that Setup is going to perform a routine check on your system, as in **Figure L3.3**. Press <Enter>.

3. Microsoft's ScanDisk utility (**Figure L3.4**) will run an abbreviated check on all hard drives installed in your system. If it discovers any problems, the installation may be aborted. If not, it will tell you that it discovered no problems. Either way, you have either the option of viewing the log or exiting ScanDisk. Press <Tab> to move over to Exit and then press <Enter>.

4. Welcome to Windows 98. Microsoft gives a sales pitch for the OS that you've already purchased (**Figure L3.5**) and over on the left side of the screen tells you that Setup will take

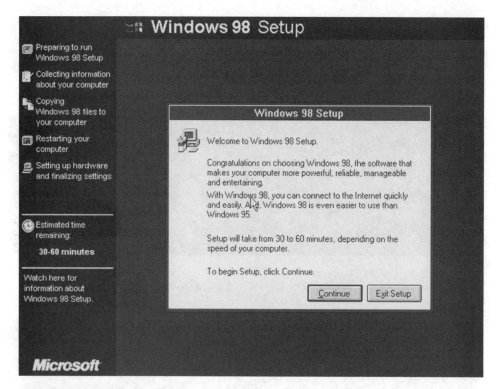

Figure L3.5 The graphical Setup

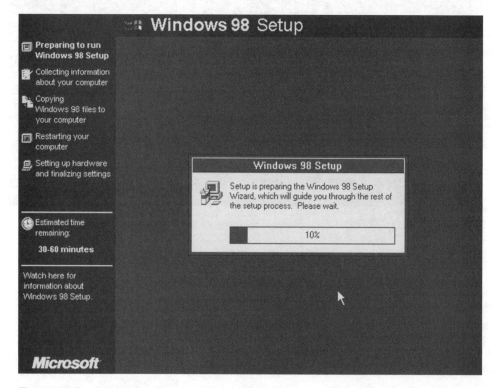

Figure L3.6 Starting the Windows Setup Wizard

anywhere from 30 to 60 minutes. It says the same thing whether you've got a 52x CD-ROM drive that can install it in about twenty minutes or a 4x that will take until your next birthday. Click Continue and Setup begins preparing the Setup Wizard (**Figure L3.6**). This will take several seconds.

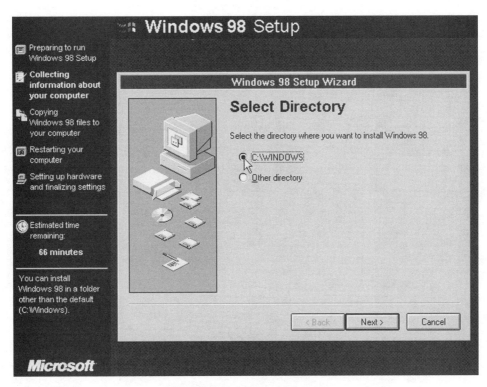

Figure L3.7 You have the option of installing Win98 to the WINDOWS directory (recommended) or to select another directory.

5. Next you will have the option of installing Windows to the C:\Windows directory or to select another directory (**Figure L3.7**). Unless you have a driving need to do otherwise (such as another version of Windows already installed on the system), select the default setting and click Next>.

6. It'll take a few moments for the directory to be prepared. But you have the lovely screen shown in **Figure L3.8** to keep you company.

7. The Setup Options screen appears next. Here, we're going to deviate a bit from a standard installation so that I can point out some interesting and very useful options that don't get installed if you select a Typical installation. Click Custom as in **Figure L3.9** and then click Next>.

8. The Windows Components screen (**Figure L3.10**) will appear. You will note that the checkboxes next to the various options appear three different ways. A clear box with a check indicates that all available options in this category have been selected. A grayed out box with a check indicates that some, but not all options have been selected. You can click the Details button to view the options that have been selected and those that haven't been. A box with no check suggests that none of the options have been selected. Click Details to see what options exist.

9. Select each category in turn to see what options haven't been selected by a Typical installation. In particular, select System Tools (**Figure L3.11**), click Details, and make sure that Backup, System Monitor, and System Resource Meter are installed. You're going to need these later on. Click OK and then Next> in the Components screen.

10. Now you need to identify your computer for the network (**Figure L3.12**). Since a later lab will involve networking these machines, now is as good a time as any to make sure the computers' names are user friendly, rather than the randomly selected collections of letters and numbers

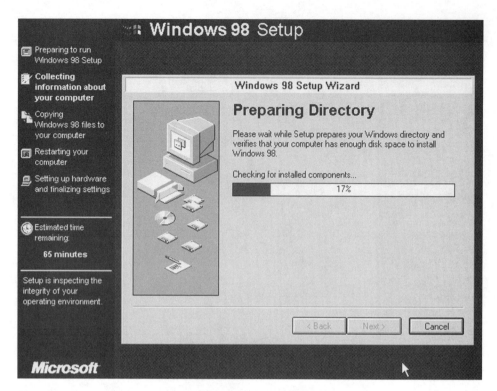

Figure L3.8 Preparing the Windows Directory

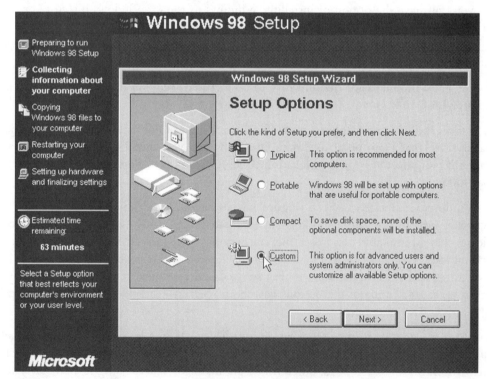

Figure L3.9 For the purposes of this lab, select a Custom installation.

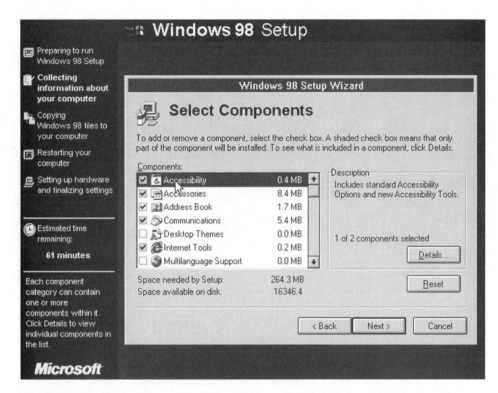

Figure L3.10   There are a number of available options in Windows that do not get installed by default.

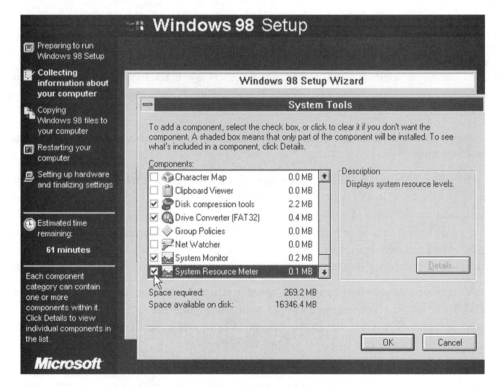

Figure L3.11   Adding new components in Windows Setup

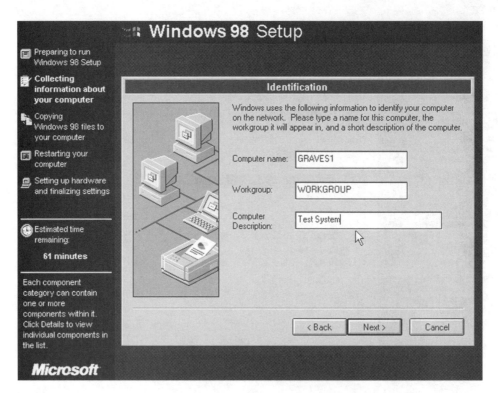

Figure L3.12   Applying computer names for networking should not be random.

Setup chooses for you. Here, the instructor should select a starting computer and each one should be named Student1, Student2, Student3, and so on and so forth until all systems have been given a unique but easily remembered name. The instructor's machine should be named HIZZONNER or HERONNER, whichever is appropriate. Click Next>.

11. On the Computer Settings screen that appears next (**Figure L3.13**), leave the defaults as they are, unless, of course, they aren't correct for your region. If not, change them accordingly and click Next>.

12. The Establishing Your Location (**Figure L3.14**) screen appears next. Select the appropriate location and click Next>.

13. Next you will see the screen titled Start Copying Files. (**Figure L3.15**). As you have most likely already figured out, this will start the file copying process. Click Next> and gather into groups to discuss the football playoffs until the process has been completed. While this is happening a number of different screens, which I have no intention of capturing for this lab, will appear, bragging about the new features of Win98. Depending on the speed of your machine, the estimated time remaining might range anywhere from half an hour to your next birthday.

14. After the file copy process has completed, your machine will reboot automatically, as shown in **Figure L3.16**. If you want to save a couple of seconds, take the CD out of the drive before the computer reboots. If not, just select the option to boot from the hard disk on the boot menu that will appear if there is a bootable CD in the drive.

15. Windows will start with the message "Getting ready to run Windows for the first time" (**Figure L3.17**) embedded in the splash screen. It will eventually move on to the User Information screen.

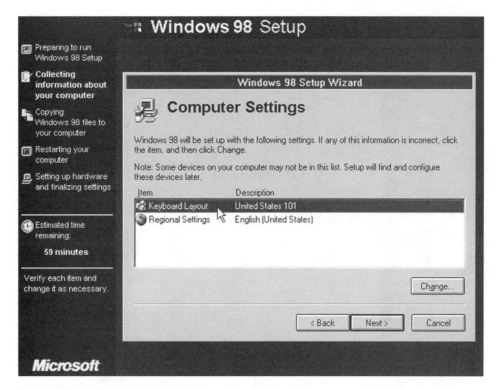

Figure L3.13   The Computer Settings screen

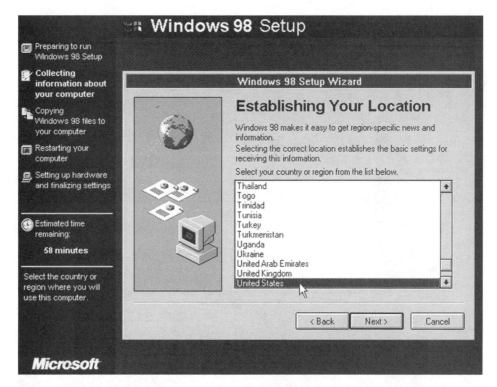

Figure L3.14   Establishing your location

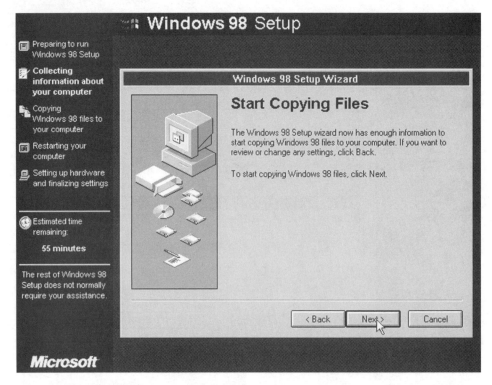

Figure L3.15   The file copy process can take anywhere from thirty minutes to eternity, depending on the speed of your machine and your caffeine level.

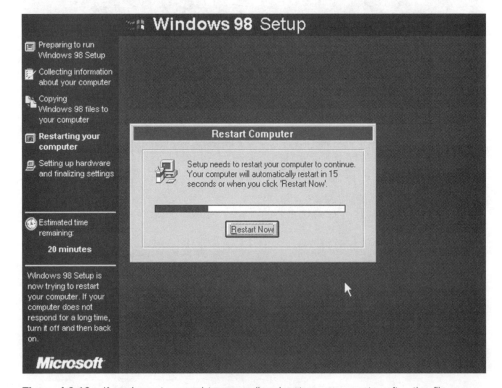

Figure L3.16   If you're not around to manually reboot your computer after the file copy process is complete, Setup will do it for you.

Figure L3.17   Starting Windows for the first time

Figure L3.18   The name is mandatory in the User Information Screen. Organization Name is optional.

16. In the window that follows, type your name and organization (**Figure L3.18**). Some sort of name is mandatory. Setup won't let you go on without one. An organization name is optional. You can leave it blank or type in The Bill Gates Fan Club if you so desire. Click Next>.

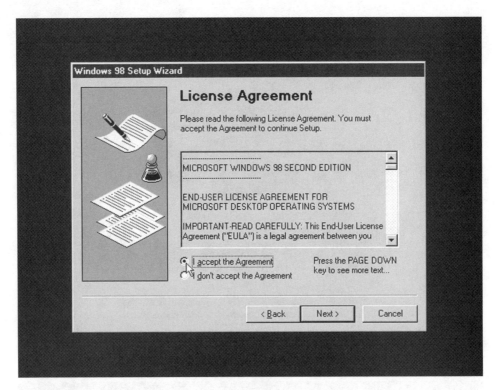

Figure L3.19   The Microsoft License Agreement, Version Win98

17. The licensing agreement appears as shown in **Figure L3.19**. Press Page Down as you read the entire agreement. Make certain you read every single word. Click I accept the Agreement and Next> to continue. If you try to be funny and decline Microsoft's generous terms, the installation will be aborted, and the rest of the class will have to wait for you until you get caught back up. Or the instructor may simply kick you out of class and tell you to come in on your own time to finish this lab.

18. The Windows Product Key screen will now appear, as in **Figure L3.20**. Type in the twenty-five letters and numbers that make up the key. It doesn't matter if you use the shift key for caps or not. Letters will be entered as capitals anyway. If you mistype any character you'll have to try again before you can continue. And don't bother trying to use the product key in the illustration. I didn't think Microsoft would take too kindly to me providing a working key in one of my illustrations. Click Next>.

19. The Start Wizard (**Figure L3.21**) tells you that it will now save all the information and that you should click Finish. That's why it's called the Start Wizard.

20. Setup now initializes the driver database (**Figure L3.22**) and scans your computer for installed hardware devices. Any Plug 'n Play device for which there is a driver in the Win98 database will be automatically installed during the next step (**Figure L3.23**). Those not recognized will need to be manually installed after the installation is completed. When the hardware scan is finished, your computer will restart once again.

21. The Setting Up Hardware screen will be the next thing you see. Setup is now installing the device drivers for the hardware it was able to detect.

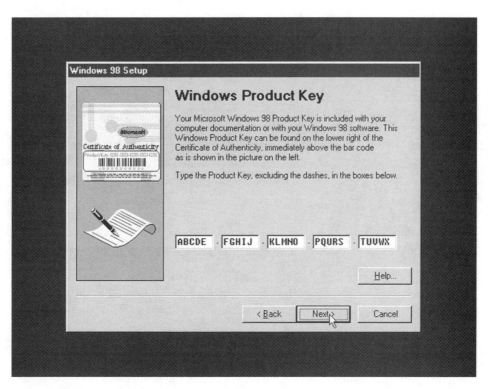

Figure L3.20 Missing even one letter in the Product Key will cause you to have to reenter it.

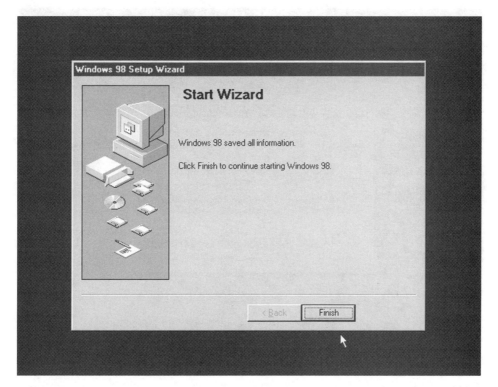

Figure L3.21 Finishing up isn't really finishing up, as you're about to find out.

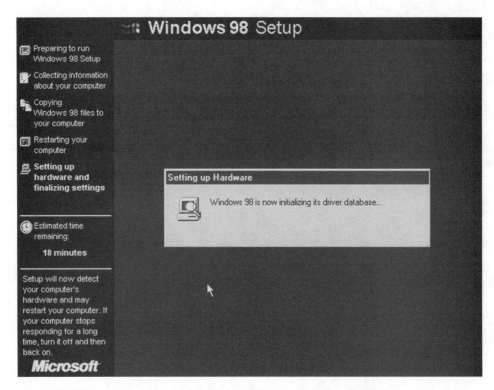

Figure L3.22 First the driver database is opened and read by Setup.

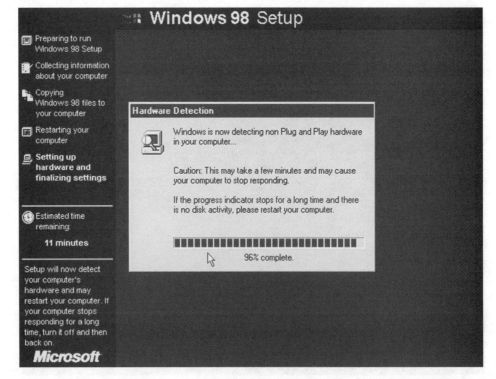

Figure L3.23 Then the Plug 'n Play devices are set up.

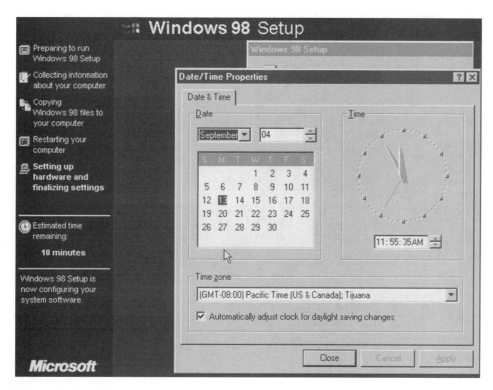

Figure L3.24 Setting the date and time

22. The Date/Time Properties (**Figure L3.24**) screen will eventually appear and you will have the opportunity to set the computer's time and date. By default the selected time zone is Pacific.

23. Now Setup will configure the Control Panel, the Start Menu, Windows Help, how Windows will handle DOS programs, and the configuration settings as shown in **Figure L3.25**.

24. When this is finished, your computer will restart yet again. As Windows starts this time (which may take a bit longer) it will attempt to install devices that weren't installed during the previous hardware installation process. This will include devices hooked up to peripheral ports such as your monitor and non-Plug 'n Play devices. For the latter, in order to successfully install these devices, you will need the device drivers on either floppy diskette or CD-ROM.

25. The Welcome to Windows 98 screen will now appear. In the lower left-hand corner of this screen, deselect the checkbox that says Sho<u>w</u> this screen each time Windows 98 starts and close the screen.

EXERCISE 1 REVIEW

1. You've just booted your computer to a Windows 98 Startup diskette and now your CD-ROM drive has just been renamed from Drive D: to Drive E:. What happened?

2. What is the first thing the Windows Setup program does after it starts?

3. In the Windows Components screen, some of the boxes are grayed out with check marks inside. What does this indicate?

4. What happens when the first phase of file copying has been completed?

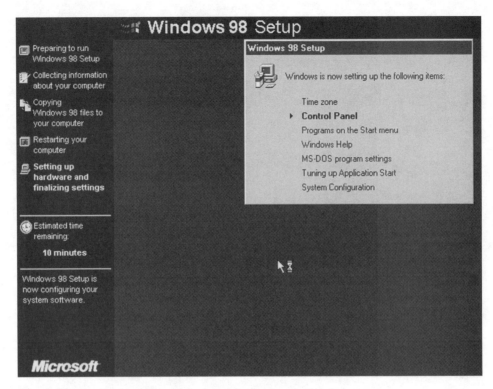

Figure L3.25   Final system configuration (well, almost, anyway).

5.   When the Licensing Agreement appears, you say out loud, "What a crock!" and click I Disagree. Now what happens?

EXERCISE 2: CREATING THE TECHNICIAN'S BOOT DISKETTE

The Windows 98 Startup Disk is fine and works quite well. But it carries a lot of excess baggage you don't need for a basic troubleshooting boot disk, and it doesn't include a couple of utilities you'll find useful. In this exercise you'll create the Technician's Boot Diskette.

1.   Open a command prompt. To do this, you can either click Start→Programs→MS-DOS Prompt, or you can click Start→Run and type **command** into the command field. If your diskette needs to be formatted before using it, you can type **format a: /s** at the command prompt.

2.   Insert a blank, formatted high-density floppy diskette into drive A: and type **sys a:** at the command prompt. Your computer will think about it for a few moments and then copy some files to the floppy and create an MBR on that floppy. The diskette is now bootable, but it isn't finished.

3.   To open Windows Explorer, right-click the Start button and select Explore. Browse to the C:\Windows\Command directory, right-click the file MSCDEX.EXE and select copy. Now right-click A: and select Paste.

4.   Browse to the C:\Windows directory and locate the files EMM386.EXE and HIMEM.SYS. Copy those file to your floppy.

5. Now, insert the original boot diskette that you used to start your original installation. You're looking for one of two files. I've seen both used from time to time. The file you're looking for on the floppy is either MTMCDAI.SYS or OAKCDROM.SYS. Right-click this file, select copy, and then paste it to the C:\Windows\Temp directory. Put the diskette for your new boot disk back in the drive and cut and paste the file from C:\Windows\Temp to A:.

6. Now, back at the command prompt, type **edit**. The MS-DOS Editor will appear. First you want to create a CONFIG.SYS file. Type the following lines, pressing the <Enter> key after each line.

```
FILES=32

BUFFERS=32

Stacks=9,256

LASTDRIVE=Z

DEVICE=HIMEM.SYS

DEVICE=EMM386.exe

DOS=HIGH

DEVICEHIGH=MTMCDAI.SYS /D:CDROM
```

7. From the Editor menu, select File→Save As and in the file name field, type **config.sys.**

8. Now, select File→New and you'll create an AUTOEXEC.BAT file. In the Editor screen type the following lines:

```
PROMPT $P$G

MSCDEX.EXE /D:CDROM
```

9. Save this file, and you're finished with this part. But there's still more work to do on this disk.

10. From the C:\Windows\Command directory copy the files EDIT.COM, FDISK.EXE, FORMAT.COM, XCOPY.EXE, XCOPY32.EXE, and XCOPY32.MOD.

11. To test your new disk, leave it in the drive and reboot your machine. When you're finished, save the diskette. You'll be using it later to create the Technician's Boot Diskette.

EXERCISE 2 REVIEW

1. What are two different ways to open a command prompt in Windows 98?

2. What is the correct command for making a formatted diskette bootable?

3. You have a diskette that needs to be formatted before you can use it to create a boot diskette. What command can you use to format the diskette and make it bootable at the same time?

4. What files will you copy from the C:\Windows directory of a Windows 98 machine to your Technician's Boot Diskette?

5. What are the two files that you need to create on your boot diskettes?

EXERCISE 3: INSTALLING THE GRAPHICS ADAPTER (OPTIONAL)

The idea of Plug 'n Play is that you're not supposed to have to install and configure any new hardware. It's a great idea when it works. The majority of the time, there are certain devices that have to be installed manually. Later in this manual, you will be installing the NIC for networking, and you'll get another chance to take a look at installing devices. Therefore, if the computers in your lab automatically detected and installed the video card, you can call it quits and move on to the next lab. The video card, on the other hand, is one of the more common culprits for *not* being auto-detected. This is especially true of cards that came out after Win98 was discontinued. If your instructor is hard-core (like me), you'll probably find yourself uninstalling and reinstalling the drivers even if the device was automatically installed. Shall we begin? You're going to love the detailed procedure.

1.   Insert the disk that shipped with the graphics adapter.

2.   Follow the instructions.

Sorry it isn't more detailed than that. Unfortunately, manufacturers of graphics adapters have more varied methods for installing their drivers than any other device. Fortunately, in this day and age, installation is so automated, they can get away with that.

LAB REVIEW

1.   You've just inserted the boot floppy into your drive and started a machine on which you wish to install Win98. You type **D:\SETUP** and get the error message "Bad command or file name." What did you do wrong?

2.   What are some of the tools you might not have at your disposal if you accept a factory default installation?

3.   What are two devices that are frequent culprits for resisting the charms of PnP?

LAB SUMMARY

Okay, you now have a workable OS on your computers. For many of you this wasn't the first time you've done this. But for those of you who have never installed an OS before, it should have been a satisfying experience. Windows 98 is a bit more cumbersome than some of the more recent Microsoft offerings, but it is a lot easier than other OSs such as Novell or Linux. As you proceed through the remaining labs in this book, you'll see what I mean.

CONFIGURING THE DESKTOP

Now that you have a working OS on your computer, how can you make it *your* OS and not just a clone of everybody else's? Microsoft was fully aware that people were going to want to personalize their computers and provided a number of ways to customize the interface. In this lab, you're going to make your computer unique. If you're working in pairs, try not to get into a fight. You can each set up your own profile so that you both enjoy personal configurations on the same machine.

A couple of short notes about later Windows versions are in order. While Win98 supports profiles, this is something that must be selected. In WIN2K and WINXP, each user account automatically carries its own profile, and no additional steps are necessary. Second, although most of the procedures in this lab are identical in the later versions, there are some minor differences in how things are applied. Rather than set up multiple labs that do basically the same thing, I will merely point out the differences in this lab.

All you need for this lab is the desktop PC onto which you installed Win98 in the previous lab. There is only one CompTIA objective covered in this lab:

1.1 Identify the major desktop components and interfaces, and their functions. Differentiate the characteristics of Windows 9x/Me, Windows NT 4.0 Workstation, Windows 2000 Professional, and Windows XP.

EXERCISE 1: CUSTOMIZING THE INTERFACE

A number of different display settings allow users to create a computing environment more pleasing to their specific tastes. Among the different tweaks that I'll be examining in this portion of the lab are the following:

- Screen resolution

- Icon size and spacing

- Icon font size

- Desktop wallpaper
- Screen saver
- Active Desktop

Exercise 1a: Setting Screen Resolution

1. Right-click the desktop to bring up the menu shown in **Figure L4.1**. Click Properties. Another way to get to this setting is to click Start→Settings→Control Panel and then double-click the Display icon.

2. This will bring up the Display Properties window as shown in **Figure L4.2**. As you can see, there are a number of different tabs at the top of this window. I'll be going over each one in turn in this lab, but this is a situation where I can say with absolute certainty that the last shall come first. Click the Settings tab. The optimal settings for resolution and color depth can vary from one monitor to another, but there are some general rules of thumb you can follow. 15" monitors fare well with 800x600 settings, while 17" monitors can handle 1024x768 (or even higher) resolutions. If your adapter supports a very high resolution along with 32-bit color, select both. In Figure L4.2, I've chosen 1024x768. Make your selections and click Apply and then OK.

3. The next window (**Figure L4.3**) warns you that applying changes without restarting the computer can result in compatibility issues. With video driver settings, this was a fairly rare occurrence, so in this case, I've chosen to make the changes without restarting. Your resolution and color depth settings have now been successfully reconfigured.

Figure L4.1 Right-clicking the desktop brings up a menu of different options specific to the display settings.

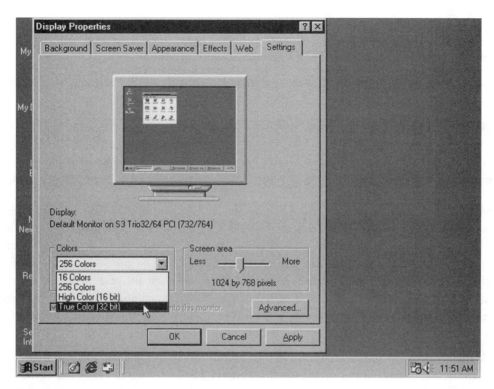

Figure L4.2 The Display Properties window of Win98

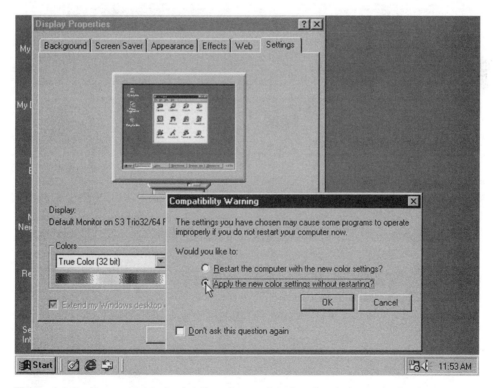

Figure L4.3 If you want to be totally safe, go ahead and restart your system.
However, Display Settings rarely cause issues, and it should be safe to accept changes
without restarting.

EXERCISE 1A REVIEW

1. What are the two ways to get to the Display Properties window?

2. Is it absolutely necessary to restart your system after adjusting display settings?

EXERCISE 1B: MANAGING ICONS

Okay, now you've got a screen with nice smooth, accurate color and virtually invisible grain. But the icons are so *tiny*! And who can read those captions? Perhaps it might be a good idea to change that.

1. Get back into Display Properties, using whichever method you prefer. Click the Appearance tab. Here is where there is a slight difference between Win98 and later Windows versions. In Win98, you will get the screen shown in **Figure L4.4**. Here you can select a preconfigured scheme that preselects all the settings I'm about to discuss, or you can individually configure specific items. What would be the fun in letting Windows do it for you? What do they know about art? You'll pick your own. In later versions of Windows, the screen shown in **Figure L4.5** appears. To configure individual items, you need to click the A<u>d</u>vanced button. This will bring up the screen that allows you to configure individual items.

2. Click the scroll arrow next to the Item list and scroll to the top. By default, you'll enter the list with an item near the bottom highlighted. In **Figure L4.6**, I've set the icon size to be 48 pixels and the font size to be 10 points. On super-high resolutions, you might want to make them

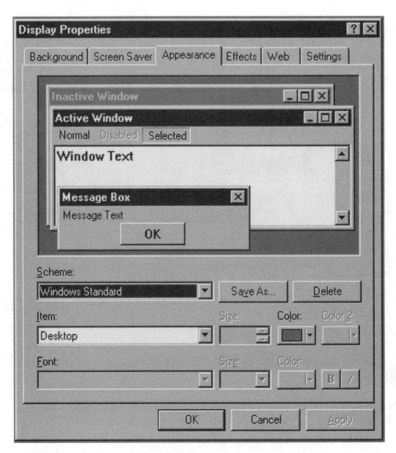

Figure L4.4 Configuring individual desktop items in Win98

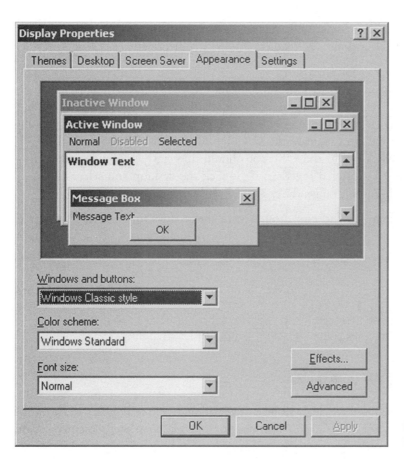

Figure L4.5 Configuring individual desktop items in later Windows versions

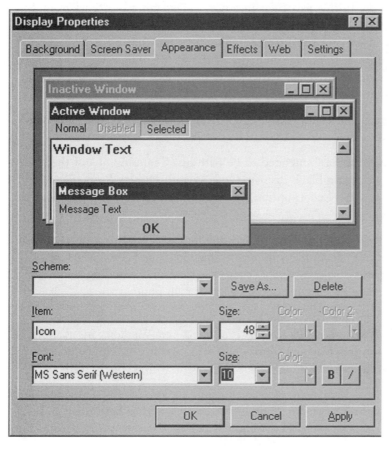

Figure L4.6 Making the icons and the font used in their captions bigger makes a high-resolution screen much easier to read.

even larger than that. If you want, you can even change the font to a different typeface, such as Times New Roman or one of the various script styles, if that is your preference. Click Apply and OK.

3. Wow! Now you've got bigger icons, and the letters are bigger but they're all scrunched together. Whatever are you going to do? You're going to fix it, that's what you're going to do! Go back into the Display Settings and click Appearance once again. This time, in the Item menu, scroll down to Icon Spacing (Horizontal) and increase the value by 50 to 75%. Do the same thing with Icon Spacing (Vertical). Click Apply and OK. You should now have a basic desktop you can live with. If the items are still clustered together, right-click the desktop and select the Arrange Icons option and select Auto-arrange. They should settle into place for you.

4. Go through each item in the Item list and play around with the settings to your heart's desire.

EXERCISE 1B REVIEW

1. Which tab in the Display Properties window houses the options for configuring the size of icons and the fonts used in their captions?

2. Which settings allow you to space the icons more evenly apart?

EXERCISE 1C: THE DESKTOP WALLPAPER

Of course, who wants to stare at a blank blue screen all day? (Even though you're *supposed* to be in one of your applications and working and not fiddling with your desktop!) People want pretty pictures. Want to set the desktop wallpaper?

1. Back to Display Properties you go. In this particular instance, the default tab of Background (**Figure L4.7**) is the one you want. You'll see two fields and two buttons. One of the fields is a list of preselected images you can use as a background, while the other allows you to decide whether you want a single instance of the image centered in the screen, multiple images spliced together and tiled, or the single image stretched across the entire screen. Trust me when I tell you, the last one is not a pleasant option.

2. Better yet, *don't* trust me. Select Bubbles and choose the Stretch option. Now click Apply and OK. Isn't this an attractive desktop?

3. Most of the images available in Win98 are intended to be patterns. Centering one of them on the screen isn't a whole lot better than a blank display. Go back into Display Properties, Background and select Bubble and the Center option. Click Apply and OK. I can see room for improvement here.

4. Go back into Display Properties→Background and select Bubble and Tile. Once more, click Apply and OK. A little better don't you think? The nice thing about the later versions of Windows is that a much larger selection of wallpaper images is available, including some relatively attractive photos. But it's very easy to use one of your own photos.

5. In the Display Properties→Background pane, click Browse. This opens a small Windows Explorer screen like the one in **Figure L4.8**. Since this is a brand new Windows installation, there won't be a lot of options on this machine from which you can choose. Notice that the only file type listed is Background Files. This can be any image file with a .BMP, .GIF, .HTM, or .JPG extension.

6. For the purposes of this exercise, browse to the Windows directory on your hard drive and select the HLPSTEP3.BMP file. Click Open. The Windows Explorer screen will close, and the

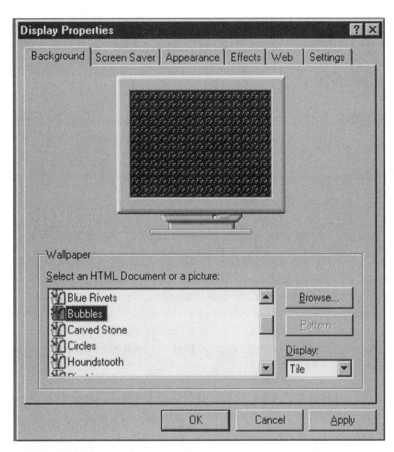

Figure L4.7   The Background tab in Display Properties

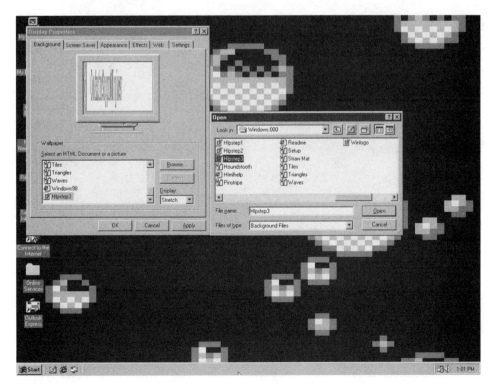

Figure L4.8   Browsing for a non-standard image file

file you just selected will be highlighted. Click Apply. You'll be told that this option is only available if Active Desktop is enabled. Do you wish to continue? You do if you want to complete this exercise. This is one of the ways to enable Active Desktop. Another way will be explored later. Click OK.

EXERCISE 1C REVIEW

1. Which tab in Display Properties allows you to configure your wallpaper?

2. How would you choose a picture of your cat to be your wallpaper?

3. Why would you **want** a picture of your cat as wallpaper?

EXERCISE 1D: CONFIGURING THE SCREENSAVER

One of the features of Windows is power management. After so many minutes, your screen will either go blank or a preconfigured screensaver will launch. Decisions, decisions! Which do you want to happen? Also, how long do you want to wait before you let your screensaver take over? As a writer (using the term in its loosest sense), I'm prone to long pauses while I ponder the next passage. The last thing I want is my screensaver kicking in just as I'm getting ready to type my next glorious phrase. Timing is everything. So now you'll configure a screensaver.

1. Yet once again, go back into Display Properties, and this time, select the Screen Saver tab. This will bring up a screen like the one in **Figure L4.9**. There are really only two actions you can

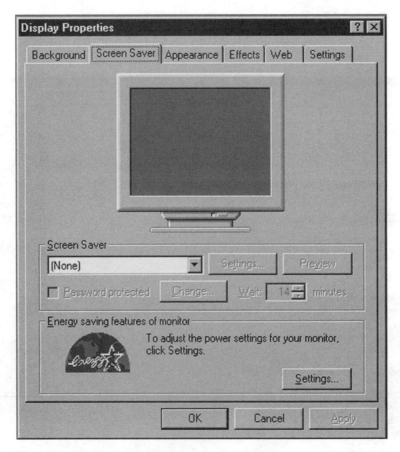

Figure L4.9 The Screen Saver screen in Windows

take here. The rest are all grayed out. One is to select which screensaver you want, and the next is to configure the energy saver settings for your monitor. The Settings, Preview, Password Protect, Change, and Wait buttons are all gray. That's because you have not yet selected a screensaver.

2. Select 3D text. Notice how the grayed out buttons suddenly become active? Click Settings to get the window shown in **Figure L4.10**. You have the option of typing in what text you want displayed or telling the system to display the time. To type in text, overwrite the default text that says OpenGL. You can choose whether you want your letters and numbers to be solidly colored or textured. And you can also select what font to use, the size, resolution, and speed at which the graphic spins. I've chosen "Master Graves" as the text in an effort to soothe my bruised ego, chosen a texture, and adjusted the resolution. The result can be seen in **Figure L4.11**.

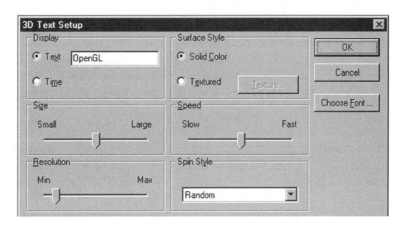

Figure L4.10   Adjusting the settings in your screensaver

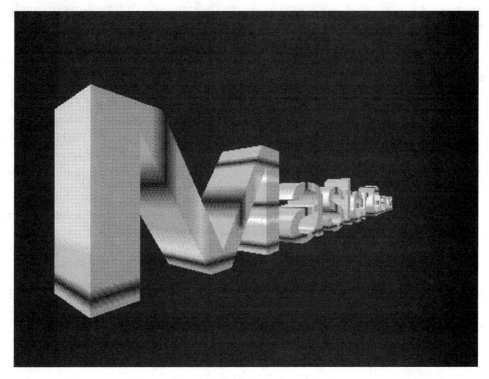

Figure L4.11   The screensaver at work

3. In the <u>W</u>ait field, type the number **1** so that you won't have to wait so long to see the results of your selection. In the real world you might want a setting that is just a trifle longer.

4. If you have materials on your computer that you'd rather other people not see, you can also configure the screensaver in such a way that after it is enabled, you can walk away from your desk. Click the <u>P</u>assword Protected box and then click <u>C</u>hange. Type in the password you want to use. Now, should unauthorized users come along and hit the spacebar on your keyboard while your screensaver is active, they'll be prompted to enter a password before they can continue. If they don't know your password, they can't browse your machine.

EXERCISE 1D REVIEW

1. Do you have to actually wait for Windows to activate your screensaver before having an idea of what it looks like?

2. Which screensaver option allows the user to put an advertising slogan up on the screen for people to read when the screensaver is engaged?

EXERCISE 2: CONFIGURING AN ACTIVE DESKTOP

Microsoft had this grand idea that if the OS and the Internet were indistinguishable, most users would be happier. After all, double-clicking is such a pain in the neck! So it made that option available. There are two more tabs in the Display Settings window that I have not yet discussed. Both are either directly or indirectly related to Active Desktop settings. Therefore, I've decided to lump them together in separate set of exercises.

EXERCISE 2A: SETTING THE ACTIVE DESKTOP

In Exercise 1c, you found out that configuring an Active Desktop was a simple matter of simply selecting a Desktop wallpaper that required it. There is also another way to get there from here.

1. Open the Display Properties windows and click the Web folder. In the screen that opens (**Figure L4.12**), simply click the checkbox labeled <u>V</u>iew my Active Desktop as a web page.

2. It's almost as simple as that. But you can get more elaborate than that. By Clicking the <u>N</u>ew button, you can connect to Microsoft's Web site and browse for Active Desktop options that are not installed from the CD during the initial installation. You can also select a specific Web page you would like displayed. For the latter, you will most likely have to know the user name and password of that site's administrator. Since it is very likely that at this point in time, most labs won't be configured for the Internet, I'll forgo demonstrating this process.

3. Now if you want to activate programs with a single mouse click on an icon there is one more step. Click the <u>F</u>older Options in the lower right-hand corner. You'll be given a warning that if you continue, Display Properties will close and the Folder Options of Windows Explorer will appear. Continue, and you'll see the screen shown in **Figure L4.13**.

4. Simply clicking the checkbox labeled <u>W</u>eb Style allows single click functionality. However, you can do even better by clicking <u>C</u>ustom (**Figure L4.14**), based on settings you choose, and then clicking the <u>S</u>ettings box. The options available here include the following:

   ■ Active Desktop

   • Enable all Web related contents on my desktop

   • Use Windows Classic desktop

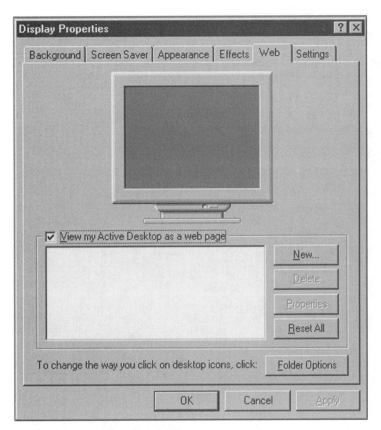

Figure L4.12
Setting the Active
Desktop

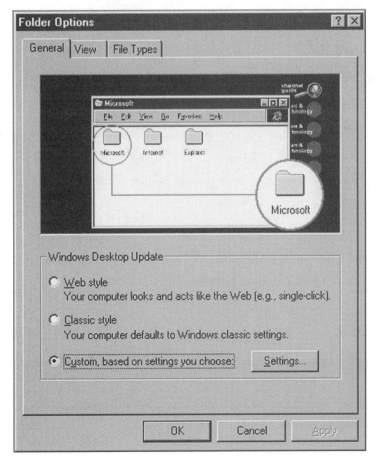

Figure L4.13
Folder Options in
Windows Exploder

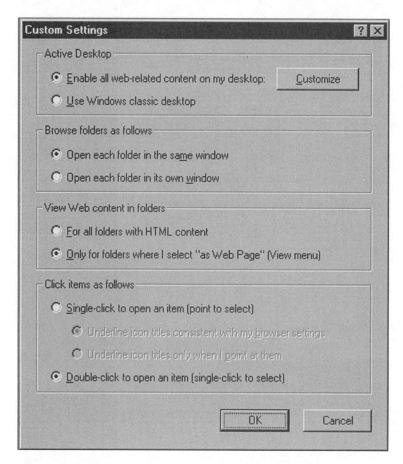

Figure L4.14 Custom Settings in Windows Exploder

■ Browse Folders as follows

- Open each folder in the same window

- Open each folder in a new window

■ View Web content in folders

- For all folders with HTML content

- Only for folders where I select "As Web page" (View Menu)

■ Click Items as follows

- Single-click to open an item (point to select)

- Double-click to open an item (single-click to select)

5. Play around with these options as time permits.

Exercise 2a Review

1. What two pieces of information might you need in order to display a Web page on your desktop?

2. How would you make sure that each item you clicked would open in a new window?

EXERCISE 2B: CONFIGURING DESKTOP EFFECTS

By now, you've already got a desktop that is pretty much unique to you, as long as everyone in class has been picking different settings along the way. Still, in Windows, you can fine-tune the desktop even further. You can configure how windows, menus, and lists behave, and, if you want, you can make sure the contents of a window remain visible even while you're dragging that window across the desktop. Allow me to show you how this is done.

1. Back to Display settings you go for yet another expedition. Click the Effects tab.

2. As you can see, you have six different options.

    a. Hide icons when the desktop is displayed as a Web page: This prevents your icons from cluttering up the Web site you've chosen to appear on your screen. It also ensures that you'll use the Start button a whole lot more.

    b. Use large icons: Umm. That make the icons on the desktop bigger. Bet I didn't have to tell you that.

    c. Show icons using all possible colors: By default, this option is selected. If your display settings are set to True Color, then the colors will blend more smoothly. Deselecting this option reverts all icons back to 256 color images.

    d. Animate windows, menus, and lists: By default this option is selected. When you click the Start menu, it sweeps open in an upward motion. Menus from the top of the screen sweep downward. For most systems this is fine. If resources such as CPU horsepower or available memory are at a minimum you can conserve by deselecting this option. The objects will simply appear when clicked.

    e. Smooth edges of screen fonts: On a low-resolution display, this reduces the stair-step effect seen on the angled edges of larger fonts.

    f. Show Window contents while dragging: When you pull a word processing document across the screen, the words and letters in the document go with the window.

3. As time permits, play around with these options. Note that reconfiguring some of them, such as smoothing screen fonts, requires the restart of the computer.

EXERCISE 2B REVIEW

1. What are the various effects that you can configure in the Windows Display settings?

2. Which of these options requires a restart of Windows before the configuration change will take effect?

LAB REVIEW

1. What happens to icons and font size when you change screen resolution from 640x480 to 1024x768?

2. Which of the Display Properties tabs would you go to in order to increase the size of icons and icon captions?

3. Where would you go to smooth the edges of screen fonts on the Windows display?

4. How could you go about putting a photo that you took on your vacation onto your desktop?

5. Why wouldn't you want to stretch the image of the brick wall to make your wallpaper?

Lab Summary

As you can see, there is a lot of tweaking you can do on the Windows desktop. You could have a dozen computers in a classroom, and no two look alike. This is just one of many ways Microsoft approached the concept of making computing a personal experience.

MANAGING PRINTERS IN WINDOWS

The process of printing in any operating system is one that baffles many. It just seems to happen. As a technician, however, you need to be able to install and configure printers, and then over the course of time, make sure that printer keeps doing what it has to do. To be certain, a lot of this work is hardware related, but the software side is every bit as critical. Updating and reinstalling drivers, mapping networked printers, and managing the print spool are all part of the job. Those are the things I'll go over in this lab.

For this lab, you'll need the lab PC with Win98 installed and, if possible, a printer with driver disks for each PC.

The following CompTIA objectives are covered in this lab:

2.1 Identify the procedures for installing Windows 9x/Me, Windows NT 4.0 Workstation, Windows 2000 Professional, and Windows XP and bringing the operating system to a basic operational level.

2.4 Identify procedures for installing/adding a device, including loading, adding, and configuring device drivers, and required software.

3.3 Recognize common operational and usability problems and determine how to resolve them.

EXERCISE 1: INSTALLING A PRINTER

This is one of those labs where it really doesn't matter what version of Windows you have installed on the system. The procedure doesn't change. Only the pretty pictures do. The biggest difference is that with Win2K or WinXP, the process of installing a local printer is so automated there is rarely much for the technician to do except supply the disk.

I should point out that installing a network printer is part of the process of managing printers. However, since you won't build your network for a couple of more labs, I've decided to include that as part of the networking lab. So for now, I'll concentrate on installing a local printer.

1. The first thing to do is to open the Printers applet from the Control Panel. There are three ways to go about this. The first is to double-click My Computer and then double-click the Printers icon. Another way is Start→Settings→Control Panel and double-clicking the Printers icon. To me, both of these methods are more cumbersome than simply clicking

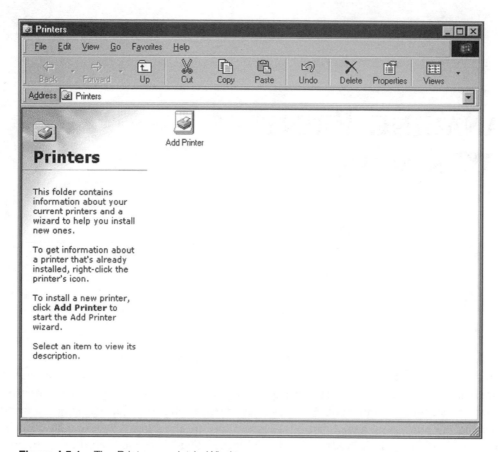

Figure L5.1   The Printers applet in Windows

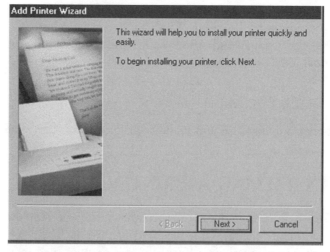

Figure L5.2   Adding a new printer is accomplished by way of a Wizard.

Start→Settings→Printers. In WinXP, there is a specific shortcut to Printers and Faxes in the Start Menu, so it's even easier. Whichever method you choose as your own, you'll end up with a screen like the one in **Figure L5.1**.

2. Because these are fresh installations, there is no printer installed, so your only option at this time is to install a new printer. Double-click the Add Printer icon to start the Wizard (**Figure L5.2**).

3. Since the next few screens are merely variations on the same theme I won't waste space or time with screen shots. Click Next and in the screen that follows, click Local printer.

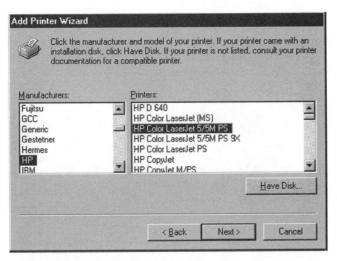

Figure L5.3   While Win98 supported a lot of printer models, it doesn't come close to the number of printers supported by XP!

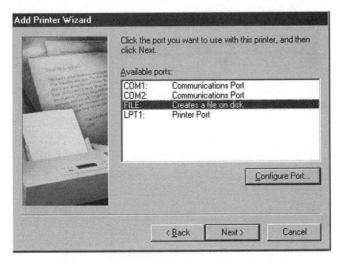

Figure L5.4   Win98 is somewhat limited in the options available for a destination port. Win2K and WinXP both allow you to print to USB and to configure a TCP/IP port from this window.

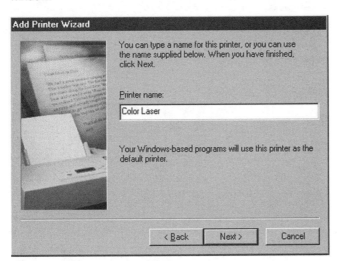

Figure L5.5   You can name your printer whatever you like. Just remember that other people will be seeing it.

4. In the next screen that appears (**Figure L5.3**) you will be presented with a long list of manufacturers supported by the OS in the left-hand pane. When you click any given manufacturer, a list of printer models appears in the right-hand pane. If you are installing a printer that is not listed, you can click the Have Disk option to install a third party driver. Since it is unlikely that most labs will be blessed with an individual printer for each student PC, you'll simply lie to Windows and tell it you have one.

5. So that everybody stays on the same page, scroll down to the HP entry in the manufacturer's list. You can get there faster by pressing the H key on your keyboard. In the printer model list, select HP Color Laserjet 5/5M PS like I did in Figure L5.3. Those of you with actual printers to install, select the make and model of printer from the list. If it isn't there, select Have Disk. Click Next>.

6. Next, the Add Printer Wizard is going to ask you what port you want your new printer to print to. In Win98 the options are somewhat limited. You have the choice of your serial ports, your parallel ports, or Print to File, as shown in **Figure L5.4**. If you have an actual printer, select LPT1. Printing to file is useful if you have a printer at the office to which you want to output the file. You can save the file the way the printer receives it to a diskette and then just run it off on any printer without having to open it up again in a word processor. Personally, I find it easier to just bring the document itself in on the diskette.

7. Now you'll be asked to name your printer. A default name will already be filled in. If you like long names, click Next >. As you can see in **Figure L5.5**, I'm going to call my virtual printer Color Laser.

8. In the next screen (**Figure L5.6**), you'll be asked whether you want to print a test page. Make sure that you have the

Win98 CD in the drive. Under most circumstances, in a new printer installation, you want to click Yes and continue. However, if there is no printer hooked up to your system, click the No button and click Next >. The Wizard will copy a bunch of files from the CD.

9. Should you ever chose to print to file, you'll get the screen shown in **Figure L5.7** asking you to name the file. After the file has been written, you'll get the screen in **Figure L5.8** asking whether the page printed successfully. Whenever you have selected a physical port to print to, Figure L5.8 will be the next screen you see. If you're printing to an actual printer, click Yes. Congratulations! Your printer is installed.

EXERCISE 1 REVIEW

1. You're trying to install a brand new printer and realize your make and model isn't on the list of supported printers. What do you do now?

2. What is the extension that allows you to identify a print job that has been printed to file?

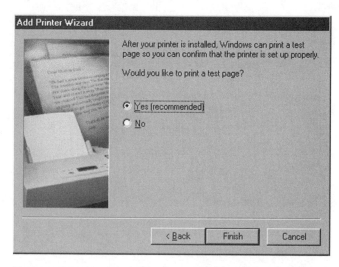

Figure L5.6 Any time you are installing a printer for the first time, it's a good idea to print a test page. That way you know for certain you have connectivity and that you've installed the correct driver.

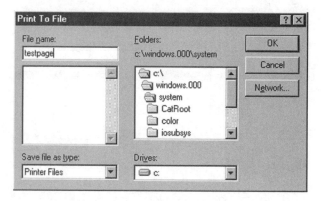

Figure L5.7 Naming a print file when printing to file is no different than in any other application.

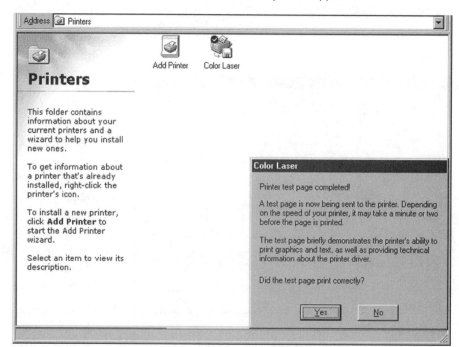

Figure L5.8 When printing to a port, the Wizard will ask you whether the test page printed successfully.

EXERCISE 2: MANAGING THE PRINTER QUEUE

One of the handier things about a graphical interface is having a utility that manages something as complex as the printer queue. What is the printer queue? Simply put, the vast majority of the time when a printer is called upon to perform, a single job gets sent from the PC to the printer, and that's that. However, from time to time things go wrong. Perhaps the printer is shared out over a network, or a user prints a barrage of documents in succession. Then suddenly the printer simply says "Enough's enough" and shuts down. The printer queue provides the user or administrator a way to cancel stalled print jobs, or even bring a critical job to the top of the list when there are numerous jobs waiting in the queue. I'll show you how.

EXERCISE 2A: REMOVING DOCUMENTS FROM THE QUEUE

In the event that a document freezes, or a networked computer locks up while printing a document, the stalled document will prevent all other documents from printing. If you're not familiar with the workings of the printer queue, this can be a perplexing problem. Everything is working properly, except the letter to your mother-in-law won't print. Remove the offending document, and all should be well.

1. The first step is to create an offending print job. With no physical printer hooked up, that ought to be easy to do. Click Start→Programs→Accessories→Wordpad. Open C:\Windows\script.doc by clicking File→Open and browsing to the Windows directory. Most likely that's the only document you will find available. It's a document file that should exist on any fresh Windows installation.

2. Press Alt+P to print the document. As the only printer installed, your Color Laserjet will be selected as the default printer. Press <Enter>. Repeat this step about four times until you have several documents in the queue.

3. Click Start→Settings→Printers to open the Printers applet (or Printers and Faxes in WinXP). Double-click the installed printer's icon to open the window shown in **Figure L5.9**.

4. All of your attempts to print will be visible in this screen. In the Window Title Bar, you will see that the printer is listed as Off Line. The information you can learn from this small applet includes the status of the print job, person who originated any particular print job (useful in a networked environment), how far along the current job is, and what time any given job was initiated.

5. To remove a single document from the queue, highlight that document and on the menu, click Document→Cancel Printing, as shown in Figure L5.9. That job will be deleted.

6. On some rare occasions the printer queue itself may lock up, or you may have to bring a printer offline for an extended period. In either case, you will most likely find it desirable to flush the printer queue completely. In order to do this, open the queue for the targeted printer, and as shown in **Figure L5.10**, click Printer→Purge Print Documents. Most frequently the

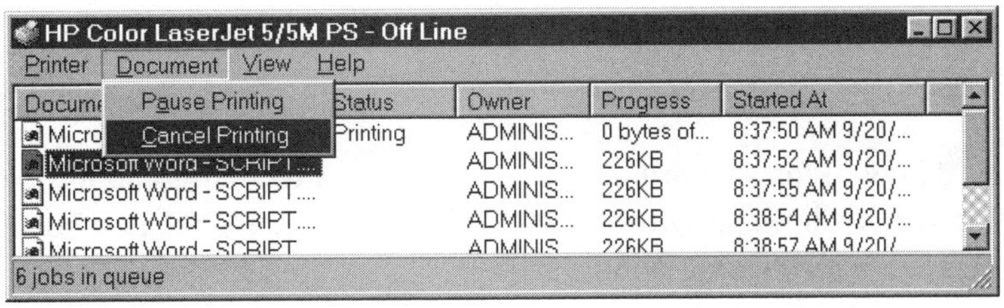

Figure L5.9   Canceling a single document from the queue

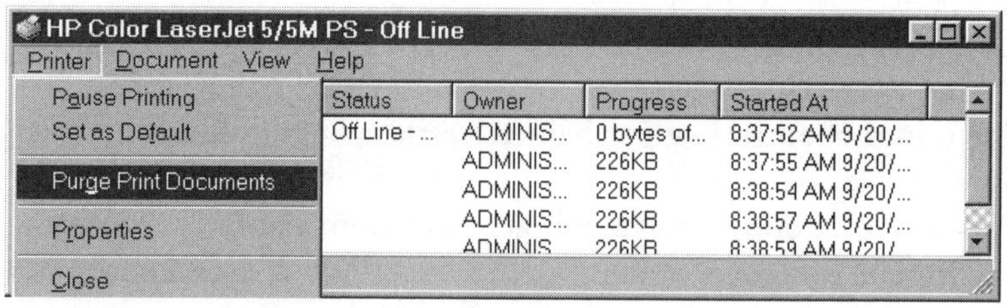

Figure L5.10 Purging the Printer Queue

document that is attempting to print will not clear itself using this method, so you would revert to the method discussed in the previous step for deleting a single print job.

EXERCISE 2A REVIEW

1. Describe the process of deleting a single job from the printer queue.

2. How would you flush all print jobs from the queue?

EXERCISE 2B MOVING A DOCUMENT TO A DIFFERENT POSITION IN THE QUEUE

In a busy office, the only printer can sometimes get quite busy. What if you have an important meeting coming up in 10 minutes and you realize you need a critical document printed. You shoot it off to the printer and realize that there are six jobs ahead of you, and two of them are copies of the boss's 800-page novel. Can you scoot your job ahead of those marathon toner-bleeders? Sure you can. Here's how.

1. Open the printer queue for the selected printer.

2. With the left mouse button, click the print job you want to move up and slide that job up underneath the job that is currently printing. Your job is now next in line. You can also move print jobs down in this same manner if you simply want to move a job to the bottom of the stack.

EXERCISE 2B REVIEW

1. What is the procedure for moving a print job up or down in the queue?

2. What happens if you keep moving the boss's print jobs to the bottom of the order?

LAB REVIEW

1. If you don't see the printer you're installing in the list provided by Windows, what option would you select?

2. Aside from LPT1, what were some of the port options available while installing a printer?

LAB SUMMARY

In this lab you learned how to install a printer and you learned a few things about managing the printer queue. This is one area that I'll revisit with WinXP. There are a couple of new tricks Microsoft taught the printer queue that are worth taking a look at.

WORKING WITH CONTROL PANEL

As I pointed out in the text book, all of the settings that make the Windows OS do what it does, and all of the adjustments that users make, are stored in the registry. However, the registry is an ungainly monster and editing it directly can be dangerous for the uninitiated. Therefore, Microsoft provided a tool for making changes to OS settings. This tool is the Control Panel. In this lab you'll take a closer look at each of those pretty little icons you see in Control Panel and learn what they all do.

The only materials you'll need for this lab are the computer onto which you installed Windows 98 and a copy of the Windows 98 CD.

There is only one CompTIA objective covered in this lab:

1.1   Identify the major desktop components and interfaces and their functions. Differentiate the characteristics of Windows 9x/Me, Windows NT 4.0 Workstation, Windows 2000 Professional, and Windows XP

EXERCISE 1: AN OVERVIEW OF THE CONTROL PANEL

Everything that makes Windows do what it does is controlled by the registry. But the registry can be a pretty scary place to be, even for the seasoned veteran. Therefore, Microsoft has provided the Control Panel for adjusting the Registry settings that are most commonly manipulated. There are two ways to get to Control Panel. One is to double-click My Computer and then double-click Control Panel. The other is to click Start→Settings→Control Panel. Either way, you wind up with a screen like the one in **Figure L6.1**.

As you can see, there are a number of icons in this screen. These are shortcuts to the various applets of Control Panel. As other services are installed onto the OS; other icons are added as well. For example, in the figure, there is an icon for HP JetAdmin. You will most likely not have that icon. I purchased a Hewlett Packard Jetdirect as a more efficient method of sharing out the printers on my network than simply hooking up a printer to somebody's computer and sharing it out. When I installed the software, the service was added to Control Panel. Also the icon with the inverted V labeled Modem Settings was added when I installed a USB modem. Other icons you're not likely to see on your screen include Administrative Tools, Adobe Gamma, and Automatic Updates. For the most part, you should see the rest of the icons in Figure L6.1 on your screen. Now I'll go over what each applet controls.

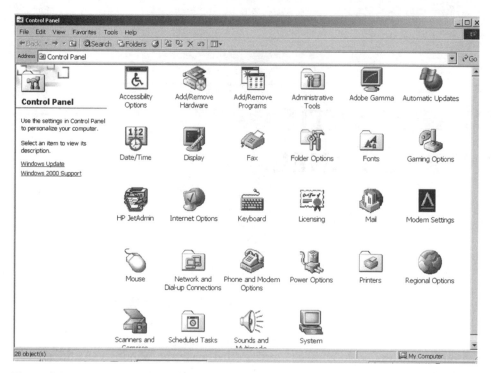

Figure L6.1 The Control Panel

EXERCISE 1A:
ACCESSIBILITY OPTIONS

Accessibility Options (**Figure L6.2**) allow you to customize the way your keyboard, display, and mouse function. Although these options are primarily designed to assist the disabled as they use a computer, many of these features are useful to people without disabilities. The various features that can be configured here include the following:

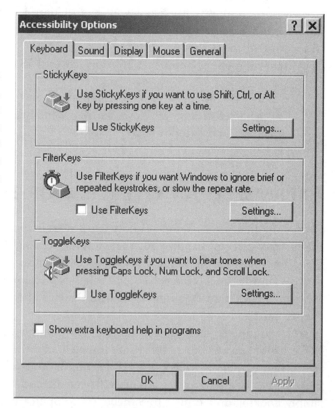

Figure L6.2 Accessibility Options

- StickyKeys: Enables simultaneous keystrokes while pressing one key at a time.

- FilterKeys: Adjusts the response of your keyboard.

- ToggleKeys: Emits sounds when certain locking keys are pressed.

- SoundSentry: Provides visual warnings for system sounds.

- ShowSounds: Instructs programs to display captions for program speech and sounds.

- High Contrast: Improves screen contrast with alternative colors and font sizes.

- MouseKeys: Enables the keyboard to perform mouse functions.

- SerialKeys: Allows the use of alternative input devices instead of a keyboard and mouse.

Set your Accessibility Options so that StickyKeys are enabled and the monitor is adjusted for High Contrast. Can you see how these options might help a person with disabilities?

EXERCISE 1B: ADD/ REMOVE HARDWARE

Figure L6.3   Add/Remove Hardware

Double-clicking this icon starts the Add/Remove Hardware Wizard (**Figure L6.3**). For the most part, Windows can automatically detect and configure any new Plug 'n Play hardware you install. Should you decide to install something that is not Plug 'n Play, or should Windows fail to install a device, the Add/ Remove Hardware Wizard comes in handy. It will do a scan of your system and detect any devices that are installed in the system. It can also detect devices that were installed but have not been properly configured to run in Windows. The Add/Remove Hardware Wizard first tries to auto-configure devices. If it fails, it leads the user through a step-by-step process of properly installing that device.

For this exercise, you're going to skip ahead a bit. At the bottom of Control Panel is another applet called System.

1.   Open the System Applet.

2.   Click the tab that says Device Manager.

3.   Click the plus sign next to the entry for Monitor.

4.   Delete the Default Monitor. Don't worry. Your screen isn't suddenly going to go away.

5.   Now go ahead and run the Add\Remove Hardware Wizard. Just follow the prompts and accept the defaults as they're offered.

EXERCISE 1C: ADD/REMOVE PROGRAMS

Back in the days when Windows 98 was considered state of the art, this applet was a lot more useful than it really is today. Still, even today, it's a very useful application. Or at least it would be if more people actually used it. It isn't the adding of programs that make it useful, but rather the removal of said programs. However, in the days of Windows 98, in order for the applet to properly remove a program, that program needed to be installed by the applet as well. When you use Add/Remove Programs (**Figure L6.4**) to install a program, Windows maintains a log of everything that is change. New folders that are created and new registry entries that are added get recorded in this log. When you decide to uninstall that program later on down the line, Add/Remove Programs uses the uninstall log to back off all changes that were made during installation. Without that log, the majority of new folders and registry entries remain on the system.

1. Make sure you have your Win98 CD available for this exercise.

2. Start the Add/Remove Programs Wizard.

3. Click the Windows Setup tab. There'll be a few moments while the Wizard examines your system for installed components.

4. Scroll down to Internet Tools, highlight it, and click Details.

5. Select Web Publishing Wizard and click OK. The Internet Tools box should now be grayed out, with a checkbox in it. Click OK. You'll be prompted to enter the Win98 CD. You may have to browse for it. It will be in the Win98 directory of the CD.

6. When the Wizard finds the file, it will copy a collection of files to your hard disk. The Publishing Wizard is now installed.

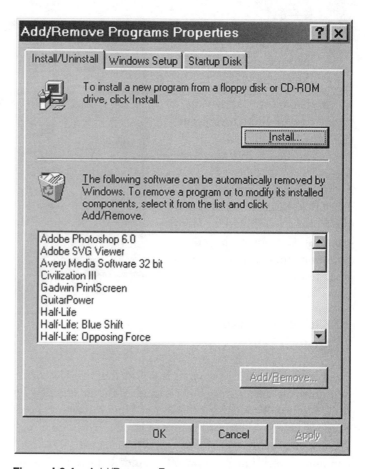

Figure L6.4 Add/Remove Programs

EXERCISE 1D: DATE AND TIME

Hopefully, the applet shown in **Figure L6.5** doesn't require a great deal of explanation. It's, um, where you set your date and time. There is, however, a third function located in this applet that you might find useful if you ever move across the country. That is the ability to change the time zone in which you reside.

1. Double-click the Date/Time icon.

2. Change the year to 1981.

3. Click Apply and then Okay.

4. Did your wardrobe suddenly get tacky and really bad music start pouring from the stereo?

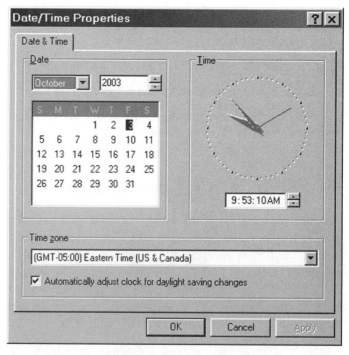

Figure L6.5 Date and Time

EXERCISE 1E: FONTS

This is another often overlooked applet in Control Panel. For most of us, our font collection grows as a result of installing new software. Any new application that provides a collection of fonts assumes you want to use everything they have to offer and installs every single one into your system. The problem with this shotgun approach is that, the more fonts you install, the longer it takes your system to load at startup. Then when you decide to use something other than your normal typeface, you have to scroll through all those hundreds of choices. In the Fonts applet (**Figure L6.6**) you can install and/or uninstall fonts to create the list useful to you. Of course, once again, like most of us, you have no idea what some of these fonts look like.

There are two ways of sorting out the wheat from the chaff. To view a font, double-click on its icon. You'll get a typeface sheet that shows the font in several different sizes. You can print that sheet out, if you so desire. Another useful trick this applet does is to sort fonts by similarity. Click View→List Fonts by Similarity. When you highlight any given font in the list, other fonts are listed as being Very Similar, Fairly Similar, or Not Similar. From what I can tell with these feeble eyes, Very Similar means Identical. If you have two dozen fonts that all show up as being Very Similar, you're better off picking one and uninstalling the rest.

1. Double-click the Fonts icon. You should see a screen that displays an icon for every font installed on your system.

2. Double-click any font and maximize the screen that appears.

3. Now click the Similarity button. The list that appears tells you how similar to the selected font every other font on the system is. Too many similar fonts tie up a lot of disk space. It's safe to eliminate them.

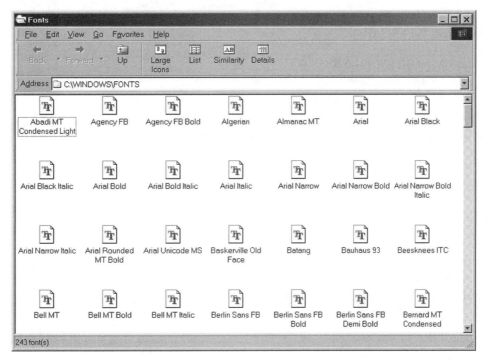

Figure L6.6 Fonts

EXERCISE 1F: INTERNET PROPERTIES

Internet Properties (**Figure L6.7**) is a very busy applet and one in which the average technician spends an inordinate amount of time. This is where you configure all the settings that let your computer successfully communicate with your Internet Service Provider (ISP). But it doesn't stop there. It's also where you can go to make sure your children aren't hopping onto websites like www.crazedmilitants.com and other sites that you'd rather them not be seeing. It is in this applet that you configure how many days worth of history to maintain and to empty the history at your convenience. To go into every setting to be found in this applet is beyond the scope of this lab, but you might want to spend a few extra moments poking around to see what there is to see.

1. Since it is unlikely that most labs will be set up for Internet at this time, simply double-click the Internet Properties icon and follow along as the instructor explains the various options.

EXERCISE 1G: KEYBOARD

The Keyboard applet (**Figure L6.8**) is, coincidentally enough, the place where certain keyboard settings are configured. These settings include the kind of keyboard installed, the repeat rate for keys that automatically repeat the character as long as the key is depressed and how fast the cursor blinks.

1. Double-click the Keyboard icon.

2. In the screen that appears, slide the bar for Repeat rate all the way up to Maximum. Do the same thing for Cursor blink rate. Click Apply and then Okay.

3. Now, since that's really annoying, put everything back the way it was.

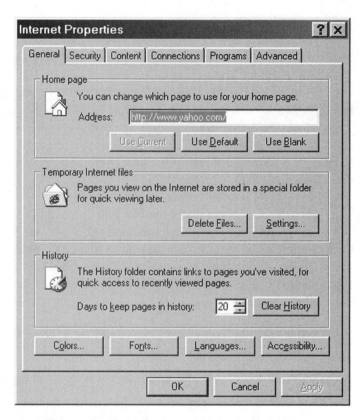

Figure L6.7 Internet Properties

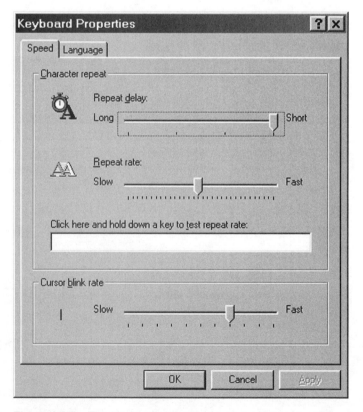

Figure L6.8 The Keyboard applet

EXERCISE 1H: MOUSE

How many times have you sat down at somebody else's computer and found out that they had set their mouse up to be left-handed? Or that the click rate was so slow that you entered the next millennium between clicks? Or so fast that Superman couldn't move his fingers fast enough to get the desired results? The Mouse applet (**Figure L6.9**) is where you change all those settings and more. In this applet you can also change the mouse cursor, enable mouse trails (not on *my* computer, you don't!) and configure how much movement of the mouse is required in order to move the cursor a certain distance. Drive your boss nuts. Make the cursor move as fast as it possibly can, set the click rate to maximum speed and then enable mouse trails. Then delete the Mouse applet from Control Panel.

1. Double-click the Mouse icon.

2. Click the Motion tab.

3. Move the slider for Pointer speed all the way down to slow and click the box for pointer trails.

4. Now since that's even more annoying than fast blinking cursors, put those settings back the way they were.

5. During break, set your partner's mouse to be left-handed, with fast cursors, and slow mouse trails. Expect him or her to do the same to you.

EXERCISE 1I: NETWORK

Another applet that you'll spend a lot of time exploring is the Network and Dial-Up Connections applet (**Figure L6.10**), especially if you pursue a career as a network administrator. This is where all local area connections for the local area network are configured in Windows 98. In Windows 2000 and XP, dial-up

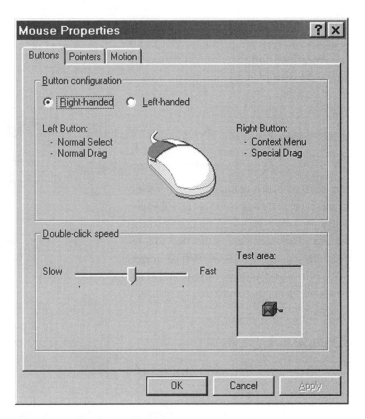

Figure L6.9 The Mouse applet

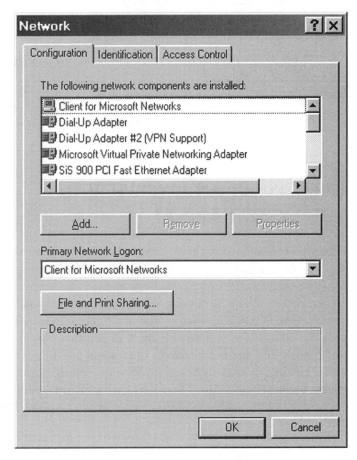

Figure L6.10 Network and Dial-Up Connections

connections were added to this applet. Because there is going to be a separate lab on networking later on, I'll pass on the guided tour at this point in time.

Exercise 1J: The System Applet

To the hardware technician, the System applet is far and away the most useful applet within Control Panel. In fact, it is so important that Exercise 2 is devoted to a review of that applet alone. Which means you're finished with this exercise and ready to move on to the next.

Exercise 1 Review

1. In terms of function, what is the purpose of Control Panel?

2. How does Add/Remove Programs make uninstalling software easier, assuming that you used the utility when you installed the program to begin with?

3. Which tab in the Internet applet allows you to configure which URL is your preferred home page?

4. Why would two different computers display two different collections of utilities in Control Panel?

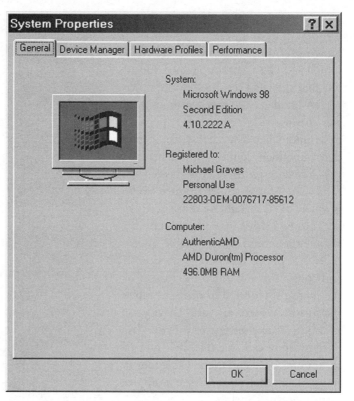

Figure L6.11 The Systems Properties screen

Exercise 2: An Overview of the System Applet

When you double-click the System icon, the System Properties window shown in **Figure L6.11** appears. As you can see, there are four tabs at the top of that screen. These tabs are General, Device Manager, Hardware Profiles, and Performance. By default, the Systems Properties screen opens to the General page. There is nothing on this screen that you can change, but it does show some good information. It tells you what version and build of Windows 98 you're using. It tells you to whom the product is registered and records the CD key that was used to install the software. It tells you how much memory there is available to the system, and it tells you the make and model of CPU running. Note the oddball amount of memory running on this particular system. This is because this particular system has on-board video. There is 512MB of RAM installed, but 16MB have been allocated to video. Since that's not available to the system, Windows reports 496MB as being available. Now take a look at the other screens.

1. Click the Device Manager tab. The screen shown in **Figure L6.12** appears. I deleted the driver files from the NIC and the sound card so that you would have some icons other than the normal ones to view. The yellow exclamation point you see next to PCI Ethernet Controller is the result of that. This indicates that it found a device installed in the system, but couldn't find the driver for that device. A red X across the icon would mean that the registry is telling Windows that a device driver has been installed for a particular device, but Windows doesn't detect the device. Hopefully, your systems won't have any warnings like these.

2. With Computer highlighted, click the Properties button. This will bring up a screen like the one in **Figure L6.13**. Here, you can view the system properties by IRQ, I/O address, or DMA channel, or view how system memory is being used. By default, Interrupt request (IRQ) appears first. If you scroll down this screen, you'll see how the fifteen available IRQs have been allocated in your system. You will also note that the same IRQ is used more than once. How this works is covered in Chapter Nine, Examining the Expansion Bus of the text.

3. Now click Input/output (I/O). You should get the screen shown in **Figure L6.14**. This shows the base I/O address for every device installed on the system along with areas of buffer memory that have been assigned to that device

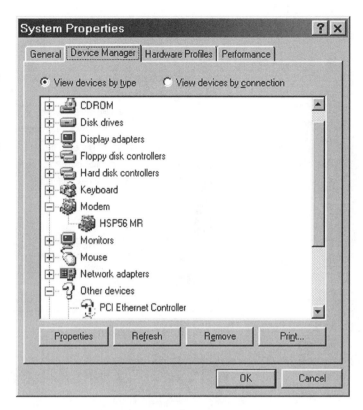

Figure L6.12   Device Manager

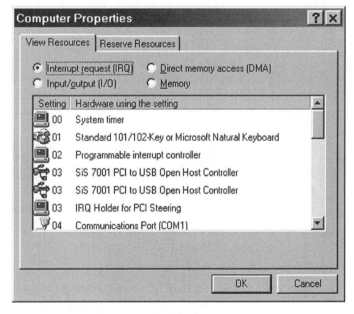

Figure L6.13   The system's IRQ allocations

by Windows. Anytime you get repeated "memory errors" that always occur at the same address, check this list to see whether that address falls within one of the ranges listed here. If so, you know what device is causing the problem.

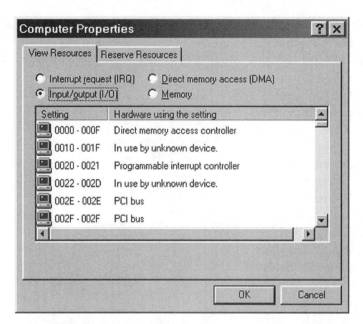

Figure L6.14   I/O properties

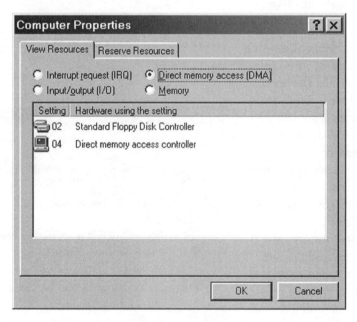

Figure L6.15   DMA properties

4.   Next, take a look at Direct Memory Access (DMA). As you can see in **Figure L6.15**, there isn't a whole lot to see here. Very few devices uses system-arbitrated DMA these days. About the only things you would see in this list are the floppy drive, a sound card, and an LPT port if it is configured to ECP.

5.   Take a look at the Memory page (**Figure L6.16**). What you see here can vary greatly from machine to machine. If your lab consists of several different makes and models of computer systems, don't expect all of them to have the same information. The numbers you see are the hexadecimal addresses of the device drivers for the devices listed alongside the address. As with I/O addresses, this can be useful in tracing those so-called memory errors Windows is always reporting.

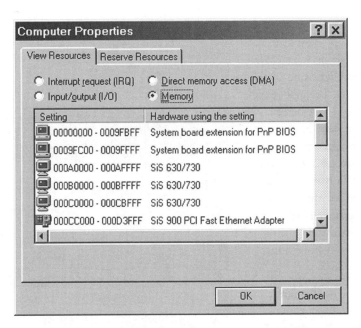

Figure L6.16 Memory properties

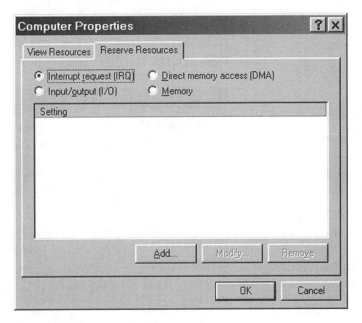

Figure L6.17 Reserving resources

6. Now for that Reserve Resources screen I've been ignoring so far. If you click that tab on any one of the properties screens you just examined, you get a blank screen like the one in **Figure L6.17**. If you click the Add button, you will be prompted to enter a value relative to the type of resource you're trying to reserve. By doing this, you are effectively taking that resource away from Plug 'n Play, and it will not be allocated to any device. This is useful if you have a device that can only be configured manually and Plug 'n Play keeps stealing the resource before your device has a chance to grab it.

7. Click Cancel to get back to the Device Manager screen. Then Click the Hardware Profiles tab. You should get the screen shown in **Figure L6.18**. Notice that, so far anyway, you only have one profile listed. It is named Original Configuration.

8. Click Copy and, in the new screen that pops up, type **New Profile**. Now go back to Device Manager and highlight one of the devices listed there. Preferably pick a device that might not be required every time you boot the machine. I have selected the modem in **Figure L6.19**. If you look toward the bottom of the Properties screen for that modem, you'll see a check box labeled Disable in this hardware profile. Check that box, as I have done.

9. Click OK. You will see the red X I described earlier in this lab. Click Close and reboot your machine. When POST has completed, as Windows is starting, you should get a boot menu offering you the option to boot to one of the following:

   ■ Original Configuration
   ■ New Profile
   ■ None of the above

10. Select New Profile and let the boot process continue. If you try to use the device you disabled, you'll be in for an unhappy surprise.

11. Finally, I will introduce you to the options found in the Performance tab of the System applet. Click that tab to get the screen shown in **Figure L6.20**.

12. Click File System to get the screen in **Figure L6.21**. I'm not going to take the time to go through each and every one of the options listed here. That is more the role of a book on operating systems. But if you like, take a look at each of the options listed to see what you could change.

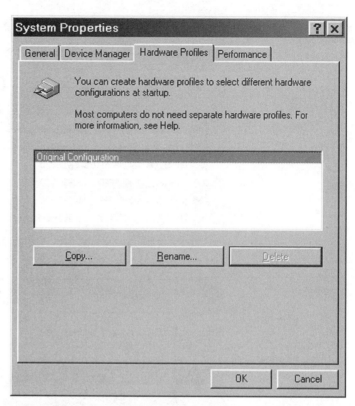

Figure L6.18   Hardware Profiles

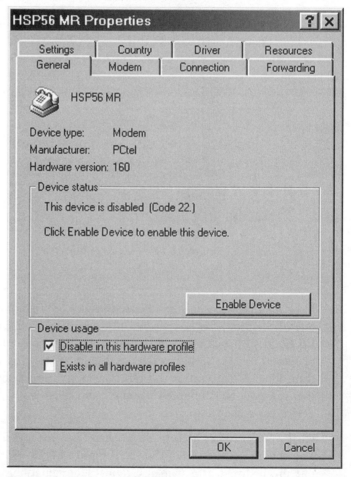

Figure L6.19   Device properties

13. Clicking the Graphics button gives you only one option—how much hardware acceleration you want to apply. Unless your system is giving you fits, the default setting of Maximum should work fine.

14. Clicking the Virtual Memory button brings up the screen shown in **Figure L6.22**. There is rarely any need to change from the default setting of Let Windows manage my virtual memory settings (Recommended). This setting adjusts the size of your swap file in Windows. Your swap file is an area of hard disk space that is treated as if it were installed memory. If your swap file is too small, system performance will drop dramatically. Therefore, if your hard disk is filling up, you might want to configure a fixed swap file so that you'll run out of disk space before your swap file is cut down to unacceptable levels.

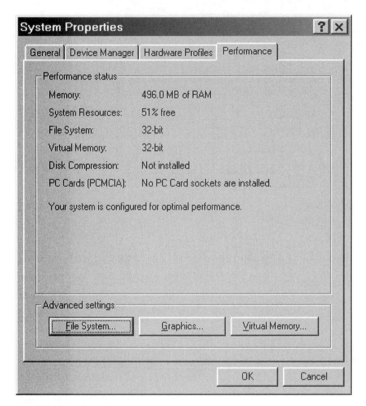

Figure L6.20   Adjusting system performance in the System applet.

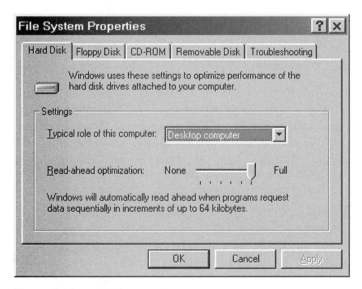

Figure L6.21   File System Properties

15. As you can see, Control Panel is a very handy program. Therefore, the last thing you're going to do is create a shortcut for it on the Desktop. To do so, double-click the My Computer icon on your desktop. Right-click the Control Panel icon and select Create Shortcut. A message will pop up warning you that you can't create a shortcut here and asking whether you want it on the desktop instead. Click OK, and the shortcut will appear on the desktop.

EXERCISE 2 REVIEW

1. What are four different views of the Computer Properties that you can bring up in Device Manager?

2. What would be the purpose of reserving specific resources in Device Manager?

3. What would the effect of reserving those resources be on the system?

4. What are two reasons you can think of for creating two or more different hardware profiles on a system?

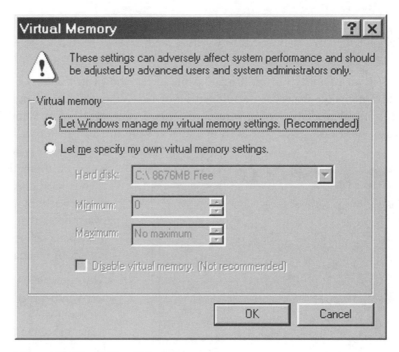

Figure L6.22   Setting Virtual Memory

5.   What would be one good reason for manually configuring the size of the swap file in Windows?

6.   What are two key functions of the System applet?

LAB REVIEW

1.   Just what is the purpose of Control Panel, and what settings are you manipulating when you make changes there?

2.   Define two different ways to access the Control Panel.

3.   What is one reason why Accessibility Options might not appear in your Control Panel?

4.   How would you create a shortcut for Control Panel on your desktop?

5.   What are four different sets of resources that can be viewed in the Computer Properties screen?

LAB SUMMARY

Control Panel is a pretty complex space, isn't it? If I didn't cover any of the icons that appear on your screen, or if I covered some that didn't appear, it's all in the way you chose to install Windows to start with. And as I pointed out earlier, adding new services will add new icons. Get to know Control Panel as well as you can, even if you're not personally a Windows user by nature. Ninety percent of your customers will be.

An Overview of the Registry Editor

As I pointed out in Lab 6 in Part 2, the Control Panel is a safe place for making changes to the registry for the most common issues. But once in a while something comes up in Windows that requires that the registry be surgically edited. Should this need arise, *be careful!* With some of the registry settings, simply having a single character missing or out of place can render your system unstable or even unbootable. Fortunately, Windows users have access to a utility that allows them to back off to a previous version of the registry. This utility varies from version to version.

All you'll need for this lab is your lab computer.

There is only one CompTIA objective covered in this lab:

1.5   Identify the major operating system utilities, their purpose, location, and available switches.

Exercise 1: An Overview of the Registry Editor

1.   The Registry Editor is one of those utilities that Microsoft has deliberately concealed from the average user. There are no pretty little icons to click on. It's a command line utility. It can be run either from a DOS window or from the Start→Run command line. In Windows 98, this utility is launched by typing the command **regedit** at either the command prompt or from the Start→Run command line. Windows 2000 and XP users have two different versions of this utility at their disposal. Typing **regedit** gets basically the same utility that Windows 98 users have. A safer, but perhaps less potent version comes from typing **regedt32** at the prompt. The latter does not have quite as powerful a search function, but it does allow the user to set security settings on individual registry keys within the editor. Since REGEDIT is available to both, this is the one I'll review.

2.   Click Start→Run and type **regedit** in the command line. You'll get the window shown in **Figure L7.1**. The left-hand window of the screen is known as the navigation area, and the right-hand window is the topic area.

3.   The Windows registry has six different keys that are stored collectively in two different files. The files are USER.DAT and SYSTEM.DAT. A third file is created optionally when System Policies are enabled. This file is called CONFIG.POL. The keys are

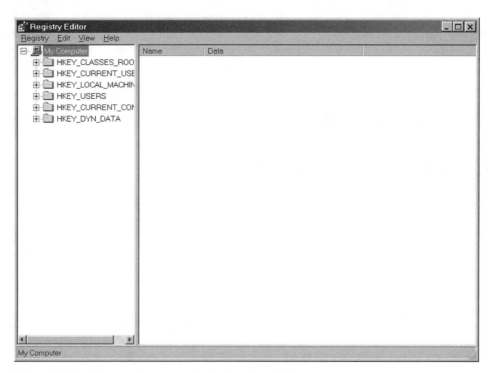

Figure L7.1   The opening screen to the Registry Editor

a. HKEY_CLASSES_ROOT: This key contains the information needed for linking objects between different applications, for determining what file types are opened by what applications, and for mapping specific functions to keystroke patterns or mouse clicks.

b. HKEY_CURRENT_USER: Here is where user-specific information is stored. This would include items such as the programs in the user's Start menu, desktop settings, applications that appear on the desktop, and display preferences.

c. HKEY_LOCAL_MACHINE: This is where information specific to the computer is stored. Device drivers, installed software, and information specific to installed hardware can be found in this key.

d. HKEY_USERS: Windows supports the ability to allow several different users to log on to the same machine, and, if so desired, each user can have individual specific settings and preferences. This hive stores the preferences and settings for all users.

e. HKEY_CURRENT_CONFIG: It is possible to set up multiple hardware profiles on the same computer. Whereas HKEY_LOCAL_MACHINE stores all hardware and software information, HKEY_CURRENT_CONFIG loads the information specific to the profile chosen during boot.

f. HKEY_DYN_DATA: This key stores dynamically configured information concerning the status of Plug 'n Play at the time of boot. Changes to device settings that do not require a reboot are changes that are managed by this key. This key is created on the fly by Windows at each startup and is not stored in any permanent file.

4. Beneath each of these primary keys are collections of hives. Hives are subkeys that contain information specific to a particular aspect of the machine. Click the + next HKEY_LOCAL_MACHINE in the navigation area to open up the hive. You should get a screen similar to **Figure L7.2**.

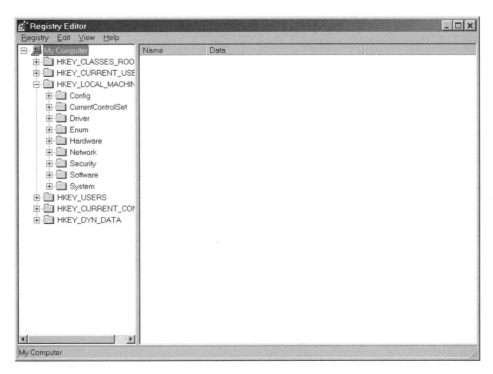

Figure L7.2   The hives of HKEY_LOCAL_MACHINE

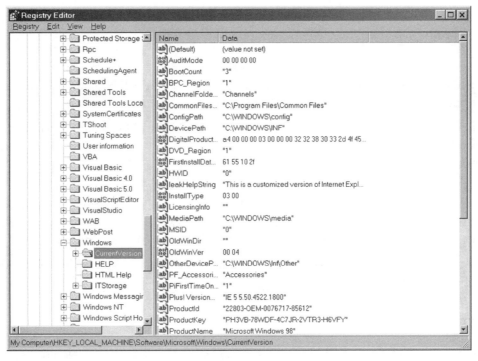

Figure L7.3   The assorted values of CurrentVersion

5. If all you did was click on the + sign, as instructed, as of yet there is still nothing in the topic area. Open the Software hive, then the Microsoft folder, and scroll all the way down to Windows. Open the Windows folder and highlight CurrentVersion. You should get a screen similar to that seen in **Figure L7.3**.

6. To see the type of data that is stored in a particular entry, double-click FirstInstallDateTime. Hmmm. They seem to have encrypted this. Wonder why they might have done that?

7. Now, so far, you haven't changed anything in the registry (or at least I hope you haven't). Before you do anything that drastic, you want to have a backup of the registry. To do that, highlight My Computer at the very top of the navigation area. Then click Registry→Export Registry File. The default location to save the file is in the My Documents folder. Accept the default folder and name your backup **REGBACKUP.REG**. Now if you screw anything up beyond recognition, you can get your system back.

EXERCISE 1 REVIEW

1. What are the two primary files that contain the registry?

2. What additional file is created if you choose to enable system policies?

3. List the six primary keys of the registry.

4. Where would settings specific to hardware installed on the system be stored?

5. Where would you look for the settings that dictate how a particular file behaves if double-clicked?

EXERCISE 2: EDITING THE REGISTRY

If you haven't made your registry backup yet, *do not continue*. Go back to the last step in Exercise 1 and complete it before going on.

1. You want to make a change that is benign and won't hurt the system. But at the same time, you want it to be something for which the results of your edit will be noticed. Right-click the Start button and click Explore. This will open Windows Explorer.

2. Click View→Folder Options. Now click the View tab. This will bring up the window shown in **Figure L7.4**. Note the line that reads "Remember each folder's view settings." You're about to change that.

3. In the registry editor, open HKEY_LOCAL_MACHINE→Software→Microsoft→ CurrentVersion→explorer→advanced folder. Highlight ClassicViewState in the navigation area. On the topic area, several entries will appear. Double-click Text. In the Edit String field that appears, type the words **These are Equal Opportunity Folders** and click OK.

4. Close and restart Windows Explorer. It is not necessary to restart the machine. Now when you view Folder Options, how does that entry read?

EXERCISE 2 REVIEW

1. How do you go about viewing system files or hidden files in Windows Explorer?

2. What was the very first thing you did before editing the registry?

EXERCISE 3: THE SCANREG UTILITY

Of course, now you want to put things back the way they were. Obviously, the easy way is to browse to that registry entry and type in the words the way they used to be. But what if you can't remember

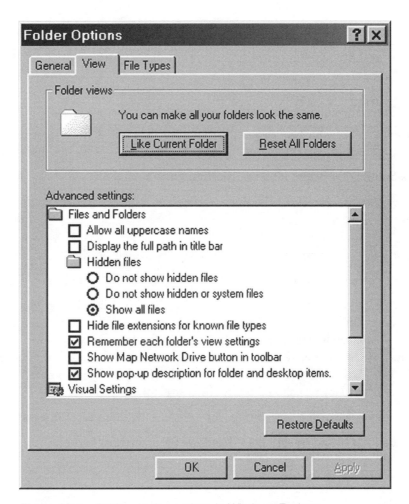

Figure L7.4   Folder viewing options in Windows Explorer

how it used to read? Or if you can't remember just what entry you edited? Or worse yet, Windows blue-screens just as it's entering the graphical mode? There's got to be a way of rolling back the system to a previous setting. And in fact, there are two ways. The first is, of course, to open the Registry Editor and click Registry→Import Registry File and then select the backup of the registry that you made. But that's too easy and doesn't give me the chance to show you the ScanReg utility. Windows 2000 and XP users, unfortunately, don't have this utility at their disposal.

1. Restart your machine, and as POST is completing, just before Windows starts to boot, start pressing the F8 key. This will bring up a boot menu. One of the options is to start Windows to a Command Prompt. Select this option and continue booting. This will bring you to a C:\ prompt.

2. Type **scanreg** at the command prompt. The ScanReg utility will start and the first thing it will do is to scan the registry for errors. You can't stop this process despite the fact that, in all the years I've used this utility, no matter how corrupted the registry was, ScanReg always reported no errors.

3. Next you are prompted to either back up your registry or to view existing backups. Obviously, if you're having a problem with the current version of the Windows registry, the last thing you want to do is make a backup of a corrupted registry. Tab over to View Backups. You will see four versions of registry backups listed by the date they were created. Pick the backup previous

to the one with today's date and select Restore. After a few moments, it will prompt you to restart your machine. Windows will boot to the older backup without the changes you made.

EXERCISE 3 REVIEW

1. Since you went out of your way to avoid backing up the registry in this particular exercise, why is backing up the registry an important function of this utility? Why not just do it in Regedit?

2. What does ScanReg do before it allows you to do anything else?

LAB REVIEW

1. Which registry key is not stored in a file, but rather is created on the fly as the system boots?

2. What are two different methods you can use to back up the full Windows registry?

3. You have just installed a new video card and during the driver installation you asked to see a list of all devices. You're so used to another computer that you use every day that you mistakenly selected a completely different make and model of video card. Now every time you boot your machine, as soon as the Windows graphical interface starts to load, the screen goes blank. How might you fix this? (There are actually a couple of ways. If you come up with a way I didn't discuss in this lab, you should lobby the instructor for extra credit.)

4. Which primary key holds the user settings for every user with an account on the system?

5. How many backups of the registry does Windows 98 maintain?

LAB SUMMARY

One of the things I hope you learned from this lab is that every aspect of how Windows performs is a function of some entry in the registry. In Lab 6 you learned how to make safe changes to the registry in Control Panel. Here I showed you the basics of how the Registry Editor can be used to make changes beyond the scope of Control Panel. And, at the risk of sounding like a skipping CD, before you mess around with the registry, *back it up!!!*

INSTALLING WIN2K PROFESSIONAL

There are those who argue that installing an operating system (OS) is not a function of a network administrator. Those are the people who either have the luxury of extremely large and diverse staffs, or simply haven't spent enough time in the real world. There will come a time when you will have to install an OS on a new system or completely rebuild an existing one.

In the following exercises, you will be installing Windows 2000 Professional.

For this lab, you will need your lab computers, a copy of Win2K Professional for each student, and four blank, formatted, high-density 3.5" floppy diskettes.

The following CompTIA objectives are covered in this lab:

1.2   Identify the names, locations, purposes, and contents of major system files.

2.1   Identify the procedures for installing Windows 9x/Me, Windows NT 4.0 Workstation, Windows 2000 Professional, and Windows XP and bringing the operating system to a basic operational level.

2.3   Identify the basic system boot sequences and boot methods, including the steps to create an emergency boot disk with utilities installed for Windows 9x/Me, Windows NT 4.0 Workstation, Windows 2000 Professional, and Windows XP.

EXERCISE 1: THE FLOPPY DISK SIDE OF WIN2K

For the most part, Microsoft has tried to forget the floppy drive ever existed. However, it has made concessions to the fact that many machines still in existence today require the services of the floppy disk drive to boot a system. If you are trying to install Win2K onto a system that is so old that it won't boot to a CD-ROM, you will need the services of a boot disk to access the Win2K CD and a set of four installation disks to install the OS. You also need a new computer.

EXERCISE 1A: PREPARING A BOOT DISKETTE

Most modern machines will easily boot from the CD-ROM drive, and since the Windows 2000 CD is a bootable CD, you would usually simply put the CD in the drive, boot the machine, and follow the yellow brick road. For the purpose of the next two exercises, you're going to attack a worst-case scenario. Your machine refuses to boot from the CD, so you must use the 4-disk set of installation diskettes.

Because this is a lab, you'll carry it a step further. Nobody knows where the setup diskettes are, so you need to make your own.

As of this writing, it is still safe to say that most network environments have a number of systems with Windows 98 installed. If this is not the case, you can download a boot image from www.mwgraves.com. To create a boot disk on a Windows 98 machine, open the Control Panel. This can be done by either right-clicking My Computer and clicking Properties and then double-clicking the Control Panel icon, or by clicking Start→Settings→Control Panel.

In Control Panel, double-click Add/Remove Programs. The right-hand tab at the top says Startup Disk. Click that tab, make sure that there is a blank, formatted, high-density floppy diskette in the drive, and click Create Disk. You now have the necessary tool for starting your computer.

NOTE: If you happen to have a machine that does not have the .cab files installed, you will be prompted for the Windows 98 CD.

EXERCISE 1A REVIEW

1. Where in Win98 is the Startup Disk creation utility located?

2. What size diskette is required to make this disk?

EXERCISE 1B: CREATING THE WINDOWS 2000 BOOT DISK SET

Using the boot diskette you just created (or were supplied), start your machine with the diskette in the drive. Select the option Start computer with CD-ROM support and let the machine boot. If your machine will not boot to the floppy, it is most likely that your CMOS simply needs to be configured accordingly. Consult with your instructor for the appropriate methods for configuring the CMOS on your particular machine

When the machine boots, it will tell you what drive letter it assigned to the CD-ROM drive. With the Win98 Startup Diskette, assuming that there is only one hard drive, this is usually E:. This is because the Startup Diskette creates a virtual drive in memory onto which it copies certain files.

From this point on, I will assume the CD-ROM to be Drive E. If this is not the case on your particular system, simply substitute the appropriate drive letter whenever Drive E is referenced. Now you're ready to begin installation.

1. Log onto the CD-ROM by typing **E:** <Enter>. At the E: prompt, type **CD bootdisk** <Enter>.

NOTE: Some schools are provided special versions of Windows 2000 for educational purposes. If you are using the 120-day Evaluation (For Educational Use) or the MSDN CD provided by Microsoft, the I386 directory may be a subdirectory buried elsewhere on the CD. If this is the case, your instructor will have the appropriate information

2. At the E:\bootdisk prompt, type **makeboot** <Enter>. The screen shown in **Figure L8.1** will appear. Have four blank, formatted, high-density floppy diskettes ready. You will be prompted to enter what drive the diskette is in. Press the A key and make sure that one of the diskettes is in Drive A. This becomes your Installation Boot Diskette. When this diskette is complete, you will be prompted to insert Diskettes 2, 3, and finally 4. To avoid confusion, label the diskettes as you go.

EXERCISE 1B REVIEW

1. How many diskettes are needed for the Win2K installation set?

2. Where is the utility that creates these disks located on the Win2K Installation CD?

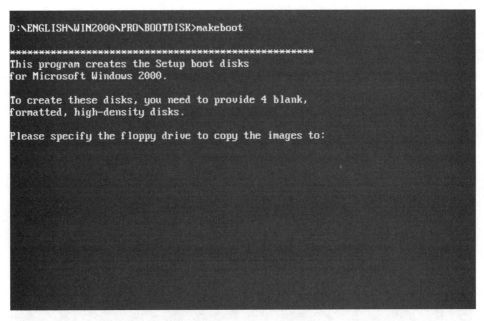

Figure L8.1   The MAKEBOOT utility is a command line utility.

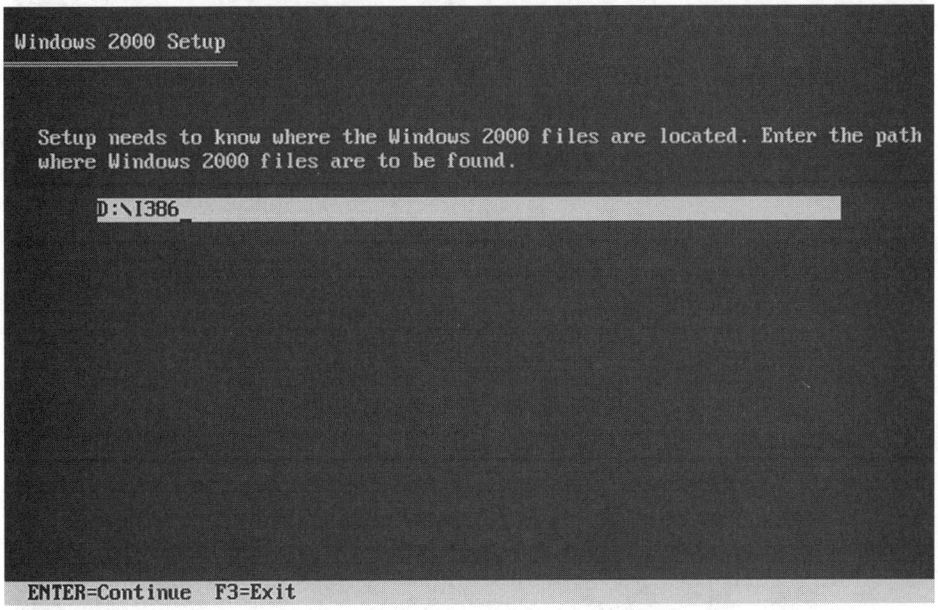

Figure L8.2   Out of the gate, the WinNT program asks you to verify the location of the source files.

EXERCISE 2: INSTALLING WINDOWS 2000

Okay, now that you've had the fun of creating those installation diskettes, you're going to love the fact that I'm not going to bother to use them in this lab. Save them for future use in your bag of tricks, but installing from diskettes is far too time consuming to fall within the constraints of a lab session. You'll boot from the CD and go from there.

1.   Place the Installation CD into your CD-ROM drive and reboot the machine. The first screen to come up is a screen that asks you to verify the location of the source files (**Figure L8.2**).

Next, a DOS screen (**Figure L8.3**) tells you that Setup is inspecting your hardware configuration. This is NTDETECT at work.

2. Next, Setup must copy some essential files to memory (**Figure L8.4**). (If you are performing this setup from the floppy diskettes set, you will be prompted for each diskette as it is needed.) This is known as the text-based portion of Setup.

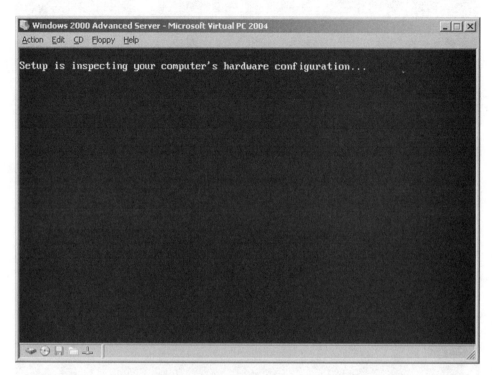

Figure L8.3 Setup's first task is to make sure you have the right hardware configuration to do the job.

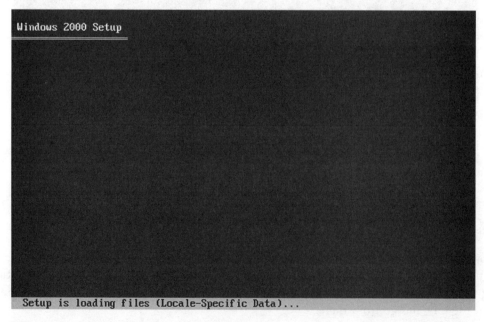

Figure L8.4 The first file copy isn't copying files to your hard disk, but rather to RAM. The hard disk hasn't been formatted yet.

3. After these files have been copied a new screen will appear, similar to the one in **Figure L8.5**, giving you three options. Pressing the <Enter> key will begin a new installation of Windows. Pressing <R> allows you to repair an existing installation using the Emergency Repair Diskette (ERD), and <F3> allows you to exit Setup without doing anything.

4. Assuming you're installing your OS to a newly installed disk drive, you'll now see the screen shown in **Figure L8.6**. You will also see this screen if you are installing over an incompatible OS such as Unix or Linux. Press <C> to continue. This will initiate the Win2K disk partitioning utility.

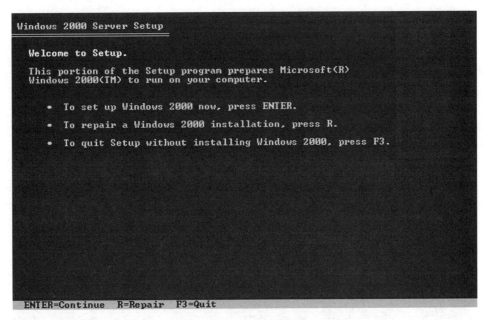

Figure L8.5   Here you have the choice of performing a new installation, repairing an existing installation, or exiting setup. Note that in this illustration, I'm installing Windows 2000 Server. Yours may say Windows 2000 Professional. The installation process is the same.

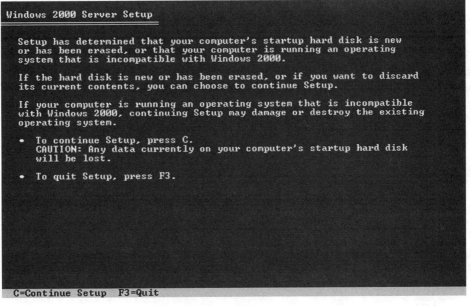

Figure L8.6   On a new or freshly FDISKed hard drive, or if you are replacing an OS such as UNIX or Linux, this window advises you that you are about to wipe your drive.

5. Now read every single word of the licensing agreement that appears in **Figure L8.7**. It's about eight pages long, so give the slower readers time to absorb all of Microsoft's generous terms. Either that, or simply press <F8> to continue. If you don't agree, this is as far as you get, and the rest of the class will have to wait until you get caught up again.

6. Now it's time to create your primary partitions (**Figure L8.8**). Because of the nature of a lab that appears later in this manual, I don't want you to prepare your drive with just one partition. For now, create a partition of 2GB, onto which the OS will be installed. Press <C> to create a new partition.

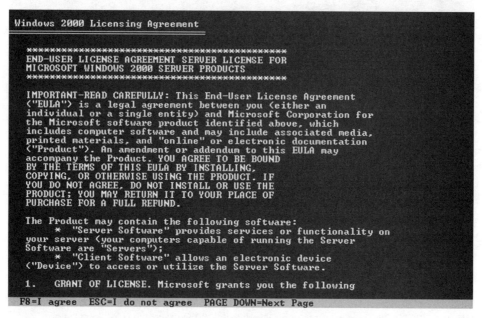

Figure L8.7   You must accept the Licensing Agreement in order to continue.

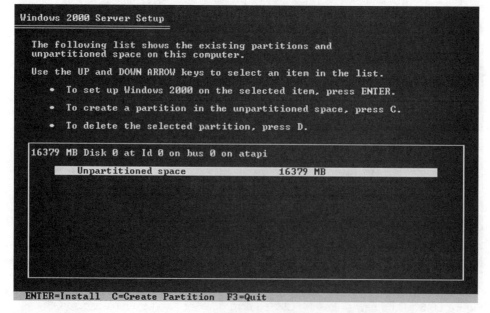

Figure L8.8   On a new drive, the first thing you must do is create the partitions.

7. In the next screen that appears (**Figure L8.9**), you will create your partitions. The default is to use the entire drive. In the bottom field, the number that is filled in represents total drive space. In order to replace that number, you must backspace to the beginning. Fill in the number **2048**.

8. When you see the next screen (**Figure L8.10**), which asks you whether you want to install your OS onto the newly created partition or into unformatted space, this may seem like a no-brainer. But in situations where you are installing a second OS onto the system, this is the point where Setup creates the information needed for your system to dual boot.

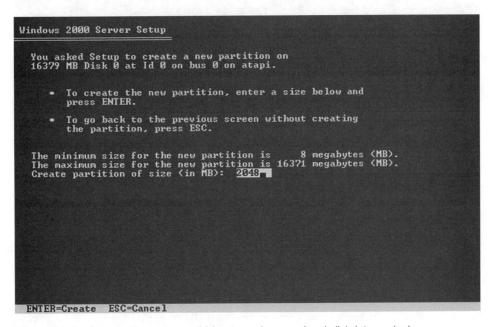

Figure L8.9 It isn't always a good idea to make your hard disk into a single partition. Put your OS and related utilities onto one partition and then install your applications and store user data on separate partitions.

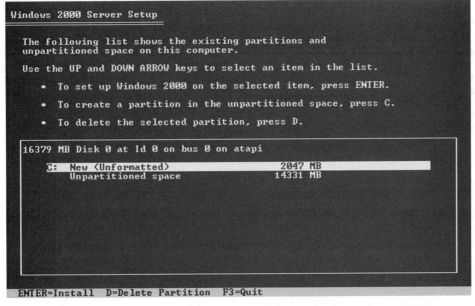

Figure L8.10 If you were installing Win2K onto a previously partitioned disk with another existing OS, this screen would identify the file system installed and how much space was occupied by that file system.

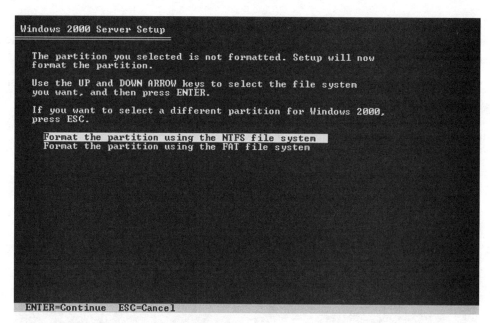

Figure L8.11 One of the bigger advantages of Win2K over previous Windows versions was the NTFS file system. Unless there is a compelling reason not to do so, Win2K computers should always be formatted to NTFS.

9. Now sit back and wait while your partition is formatted. This generally takes long enough for a cup of coffee and a donut while the class mingles in the break room discussing the football playoffs.

> **NOTE:** During the initial formatting process, the hard drive will be formatted to FAT, even if you selected NTFS as your file system of choice. If you are creating a file system of 2GB or larger, it will automatically format the drive to FAT32. Any partition smaller than 2GB will be formatted in FAT16.

10. The next option (**Figure L8.11**) allows you to choose the file system you want to use on your new partition. Your choices are FAT or NTFS. FAT is a poor choice for many reasons. The most important reason to you is that if you **don't** choose NTFS, you won't be able to do some of the later labs.

11. On reboot, you enter the graphical portion of the installation process. A lovely screen with the Windows 2000 Professional logo will appear. At the bottom a progress bar labeled Starting up will appear.

12. After that, Setup will convert the drive to NTFS. This takes a minute or two. The computer will automatically reboot once again.

13. Setup will now perform a thorough Plug 'n Play (PnP) scan (**Figure L8.12**), looking for any PnP devices and/or legacy devices previously installed on a prior OS (in the case of an upgrade). If your screen appears to flicker and Setup halts for a few minutes, this is normal. If Setup halts for an abnormally long time, it is probably hung, and you will need to restart your machine once again.

14. When this is finished, you'll have the opportunity to configure your regional settings (**Figure L8.13**). Generally in the United States no changes need to be made here. For overseas users, it might be necessary to click the Customize button. Click Next and in the screen shown in **Figure L8.14**, type your name and organization (organization is optional.)

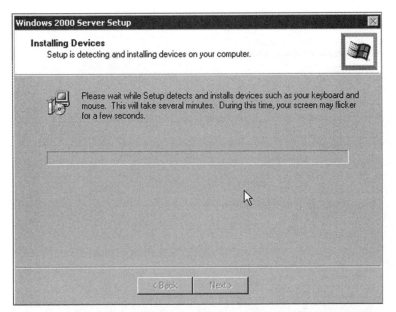

Figure L8.12 The Plug 'n Play scan detects your hardware and creates a list of device drivers to be installed. Contrary to what the screen says, not all devices are being installed at this point. But you do have mouse support now.

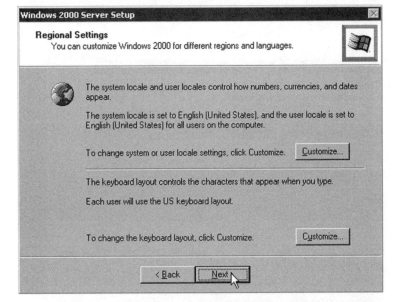

Figure L8.13 In the United States, it isn't generally necessary to make any changes to the Regional settings, but overseas users can change the currency settings as well as how numbers and dates are displayed.

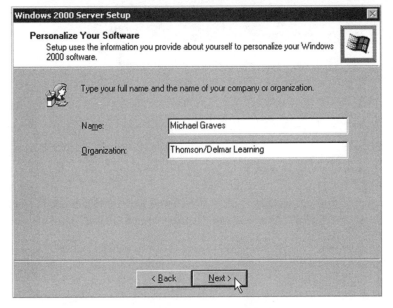

Figure L8.14 It isn't necessary to type an organization name, but Setup won't continue until you type a user name.

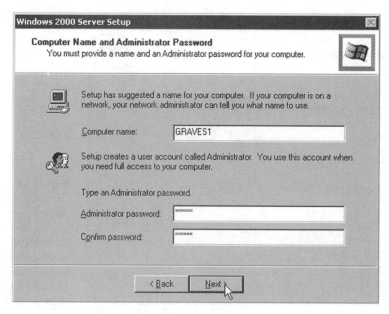

Figure L8.15 The computer name selected at random by the system isn't usually the best choice.

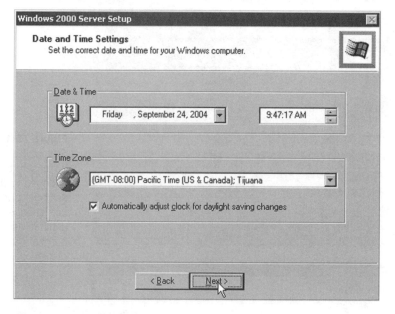

Figure L8.16 Now you know what time and day it was when I wrote this lab. Send me a present on the anniversary.

15. Now it is time to type in the twenty-five-digit CD key that shipped with the software. Type carefully, or you'll be doing it again.

16. The following screen (**Figure L8.15**) will provide a suggested NetBIOS computer name. You won't be using their suggestion. Student machines will be named STUDENT1 through STUDENT12 (or however many student machines there are in the classroom). Here is where the password is selected. For the password *all* students will simply use *password*. This will avoid the inevitable confusion when someone says, "I forgot what I used."

17. Next you set the time and date (**Figure L8.16**), if it is incorrect, and reset the time zone to your own. It will default to Pacific Standard.

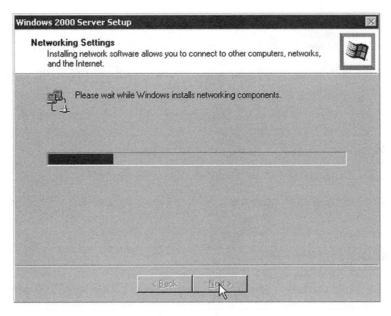

Figure L8.17   Installing networking components

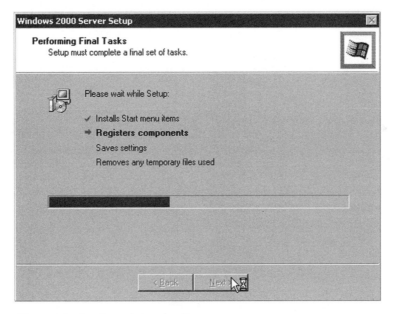

Figure L8.18   Completing the Setup

18. Setup will now begin installing networking components, followed by Windows 2000 compo-
    nents (**Figure L6.17**). Let everything install as per default.

19. In the final step (**Figure L8.18**), Setup installs Start menu items, registers components, saves
    settings, and finally, removes any temporary files used.

20. When your computer reboots once again, if there are any hardware devices that were detected
    for which Windows did not have a correct device driver in its database, you will be prompted
    to search for the appropriate drivers. At this point the drivers can be anywhere, including the
    network, if networking was configured for DHCP and there is a DHCP server available.

EXERCISE 2 REVIEW

1. What were the two choices of file system offered during the Disk Preparation sequence?

2. At what point did you have control over your mouse during the installation procedure?

LAB REVIEW

1. How many floppies are needed to create the Win2K Installation set?

2. In what directory is the utility for creating boot diskettes located?

3. What happens if you reject the licensing agreement from Microsoft?

LAB SUMMARY

The previous pages led you through a step-by-step procedure for creating all the necessary floppy disks you may ever need to install Win2K. You also spent a large portion of your day installing the product. Now you know why network administrators were so happy Microsoft created methods by which large numbers of machines could be configured at once over a network.

THE WIN2K INTERFACE

I'm going to let you enjoy one brief interlude with Win2K to introduce you to some of the interface changes. Since there are still so many computers out there in the real world running Win2K, it's a good idea if you know where to find things. What I'll be doing in the following exercises is pointing out differences between Win9x (which you just spent a decent amount of time exploring) and Win2K. It's a classic case of, "They're exactly the same, only different."

The only materials you'll need for this lab is your student lab computer, which should now be running Win2K.

The CompTIA objectives covered in this lab include the following:

1.1 Identify the major desktop components and interfaces and their functions. Differentiate the characteristics of Windows 9x/Me, Windows NT 4.0 Workstation, Windows 2000 Professional, and Windows XP.

1.4 Identify basic concepts and procedures for creating, viewing, and managing disks, directories, and files. This includes procedures for changing file attributes and the ramifications of those changes (for example, security issues).

2.4 Identify procedures for installing/adding a device, including loading, adding, and configuring device drivers and required software.

EXERCISE 1: THE START MENU

One of the ways Microsoft made Win2K a bit more user-friendly was to provide a number of different ways in which each user could configure the Start menu to his or her own taste. In the next few pages, I'll show you several different ways to customize the Start menu.

EXERCISE 1A: CONFIGURING THE START MENU

The simple Start menu is the home of some significant changes between Win9x and Win2K. These changes are very subtle, and in fact, cannot be see at first glance. They become available only when configured by the user. In the following pages, you will get acquainted with the Win2K Start menu and learn how to customize it to your taste.

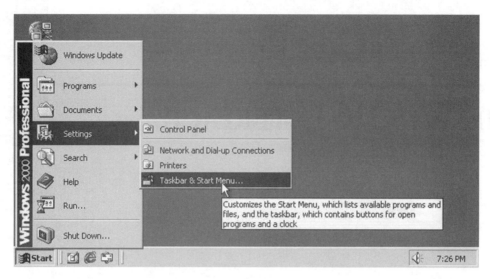

Figure L9.1   Hovering over a menu item in Win2K brings up a description of the item's properties without requiring any further action on the part of the user.

1. Click Start to view the Start menu. Examine the menu closely to see any differences between the Win9x menu and Win2K. There really aren't any are there?

2. Click on Settings→Taskbar and Start menu. (*Note:* You can also get to the Taskbar and Start menu configuration console by right-clicking an unoccupied space on the taskbar and selecting Properties.) If you allow the mouse cursor to hover over the menu selection for a second, you will see a significant change in the user interface. As you can see in **Figure L9.1,** this action brings up a description of the item's properties. Go ahead and click Taskbar and Start menu.

3. This will bring up the Taskbar and Start Menu Properties dialog box. Notice that in this window there are two tabs. The General tab is the one that opens by default. There are five checkboxes available, three of which will be checked by default:

    a. Always on top: Checked by default. Checking this box ensures that even when other programs are running, the taskbar and Start menu are not obscured by any application.

    b. Auto hide: Not checked by default. When checked, the task bar and Start menu disappear from view. When the mouse cursor is pointed to the area where the task bar resides, it will appear.

    c. Show small icons in Start menu: Not checked by default. The pictures used in the Start menu are reduced significantly in size.

    d. Show clock: Checked by default. The time is shown in the lower right-hand corner. Hovering the mouse cursor over the time will also display the date.

    e. Use Personalized Menus: Checked by default. Programs that have not been used recently are not displayed. In order to see the full list of programs, click the down arrows at the bottom of the menu.

4. Click Auto Hide to select it and Show clock to deselect that item. Click Apply and watch what happens. Click Cancel, and the desktop reverts to the way it was.

5. Now click the Advanced tab. This will bring up the screen shown in **Figure L9.2**. There are four buttons on the right-hand side of the window clustered toward the top:

    a. Add: This allows you to add items to the Start menu that do not appear by default. Just for fun, I'll have you do that in Lab 1b.

    b. Remove: I doubt that I need to explain this, but this option allows you to remove items from the Start menu that were added during the OS installation or during the installation of another application. Since applications frequently add about a gazillion items to your Start menu that you'll never use, this is a handy feature.

c. Advanced: This is actually just another way of adding or removing an item from the Start menu. But instead of using a browse feature, it opens Windows Explorer, and you locate the item you want from there.

d. Re-sort: Here is where you can rearrange the order of items displayed on the Start menu.

6. Beneath these four buttons, discretely distanced by a small space, is a fifth button labeled Clear. Clicking this button wipes out the recent history of opened documents, web pages, or programs.

7. The bottom pane in the Taskbar and Start Menu Properties window is one of the most useful. In a new installation of Win2K the various options offered here are unchecked. By selecting them, you can add the following items to your Start menu:

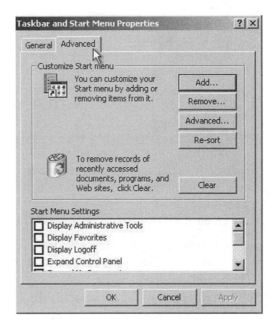

Figure L9.2 The Advanced tab of the Task bar and Start menu configuration screen allows a number of detailed changes.

a. Clicking the Display Administrative Tools box adds a shortcut to the administrative tools to the Start→Programs section of the Start menu.

b. Display Favorites places a shortcut with an itemized submenu of all your Internet Explorer favorite sites.

c. Display Logoff puts an item into the Start menu that allows one user to log off and another to log on without having to restart the computer.

d. I like this one a lot. One of the default items in the Start menu is a shortcut to Control Panel. Clicking that shortcut opens Control Panel. (What a surprise!) The Expand Control Panel option places a submenu in the Control Panel shortcut that allows the user to go directly to one of the individual applets in Control Panel without having to open Control Panel first.

e. Expand My Documents performs that same function for the My Documents entry in the Start menu.

f. Expand Network and Dialup Connections creates a submenu to the Control Panel option that provides shortcuts to each individual DUN or LAN connection configured on the computer.

g. Expand Printers provides shortcuts to each printer installed on the system.

h. Scroll the Programs Menu changes the Start→Programs section of the Start menu from a column to a scroll-down field.

8. Click Expand Control Panel; then click Apply and then OK. Now watch what happens when you Click Start→Settings and then hover the mouse cursor over Control Panel.

REVIEW OF EXERCISE 1A

1. What were the various options located in the General tab of the Taskbar and Start Menu Properties window?

2. What are two different ways to add an item to the Start menu?

3. You've just spent your lunch break searching the Internet for a new job, and you don't want a coworker who shares the computer with you to know what you're doing. How can you hide your tracks from casual eyes?

Exercise 1b: Adding an Item to the Start Menu

In this exercise, I'm going to have you add a program to the Start menu that does not install by default. A handy little program for the administrator is the Microsoft Management Console (MMC). This program is useful for creating your own customized administrative tools, but by default it is only a command-line function. You'll create a Start menu shortcut for it.

1. Open the Taskbar and Start Menu Properties window and click Advanced.

2. Clicking the Add button will bring up the window shown in **Figure L9.3**. If you know the exact path to the program you want to add, simply type it into the field labeled Type the location of the item. Because everyone knows where MMC.EXE is located, type it and click Next. What? You *don't* know where MMC.EXE is? Neither do I, so go ahead and click Browse....

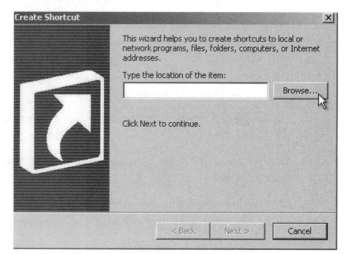

Figure L9.3 The Browse function of the Add button allows you to search your hard drive for the program you want to add.

3. In the window that opens (**Figure L9.4**) you will see a screen that displays My Documents, My Computer, My Network Places, and Connect to the Internet. Each one of these represents a shortcut that you can add to the Start menu to make your computing life easier. Adding a document creates a shortcut to the document that automatically launches the default program to edit that type of document. Adding an Internet location creates a shortcut that will launch Internet Explorer and automatically browse to that URL. You're going to add an executable, so, as shown in Figure L9.4, click the + symbol next to Local Disk C to open it. Click the + symbol next to the System32 subdirectory and scroll down to MMC.EXE. Life will be much easier and this lab will go faster if you click the first file you see under System32 and press the M key. That will highlight the first file with a name beginning with M. and scrolling will be much faster. Double-click MMC. and the file will appear in the location field. Click Next.

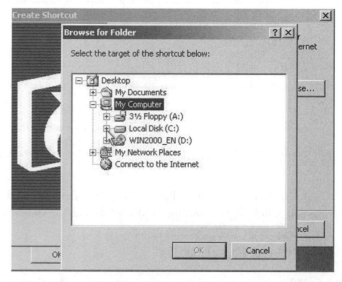

Figure L9.4 The Browse function allows the user to locate shortcut targets without having to know the exact location.

4. In the next screen (**Figure L9.5**) you will be prompted to select which folder should house the new shortcut. The default is Program Files, but there is also an option for a New F̲older. Click that button to automatically create a new folder with the same name as the shortcut.

5. Now a screen like the one in **Figure L9.6** pops up, prompting you to give your new shortcut a name. The default is the name of the executable. This might be occasionally useful, but if you happen to be creating a shortcut to a program whose executable is WXKV785.EXE, that might not be so useful to other people using your computer. Or you, for that matter, three days after you create the shortcut. Call your new shortcut Microsoft Management Console.

6. Click Finish and then click OK in the Taskbar and Start Menu Properties screen. Your new item now shows up in the Start Menu.

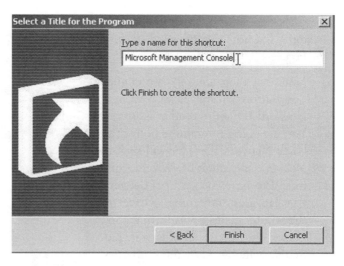

Figure L9.5   One of the options in creating shortcuts is to put each shortcut in its own folder.

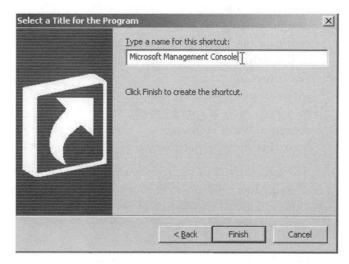

Figure L9.6   Make sure you give your shortcut a name you can recognize later.

EXERCISE 1B REVIEW

1. What is the purpose of adding new items to the Start menu? Aren't all installed applications automatically added?

2. Why is it not a good idea to accept the default name for a new shortcut?

EXERCISE 2: NAMING A COMPUTER IN WIN2K

In Win9x, naming a computer was done in the Network applet at the time networking was set up. If a network card is detected during installation, naming is done during installation. Microsoft made it slightly different in Win2K.

1. The first thing you have to do is open the System applet in Control Panel. There are three ways to do that. One is to click Start→Settings→Control Panel and then double-click System. A second way is to double-click the My Computer icon on the desktop, double-click Control Panel, and then double-click System. Far and away the easiest method is to simply right-click My Computer and select P̲roperties. Any of these methods will bring up the screen shown in **Figure L9.7**.

2. Click the Network Identification tab at the top. That brings up the window shown in **Figure L9.8**.

3. Note that there are only two configuration buttons on this panel. One is Network ID and the other is Properties. You are going to examine both of them. Click Network ID. This will start the Network Identification Wizard. Since I intend to have you set up a network in a later lab, I am not going to have you run the wizard at this point in time. So Cancel the Wizard and, back in the System window, click the Properties button. That will bring up the screen shown in **Figure L9.9**.

4. The top field in this panel is for the computer name. Here is where you configure the NetBIOS name for the computer you are configuring. There can be no two computers on the network with the same NetBIOS name. A complete discussion of computer naming conventions can be found in the text book. Fill in a computer name.

5. Beneath that and to the right is a button labeled More... I'll get to that one in a minute. Toward the bottom of the panel are two more fields in a box labeled Member of. By default the one with the Domain checkbox is deselected and the field is grayed out. The one labeled Workgroup is checked and the word WORKGROUP is already filled in. In a workgroup setup, all computers that are going to talk to one another must be a member of the same workgroup. Were you to set up a network in which half the computers in the classroom were part of the CLASSONE workgroup and the other half were part of the CLASSTWO workgroup, even though they were part of the same physical network, CLASSONE computers would not be able to communicate with CLASSTWO computers and vice versa.

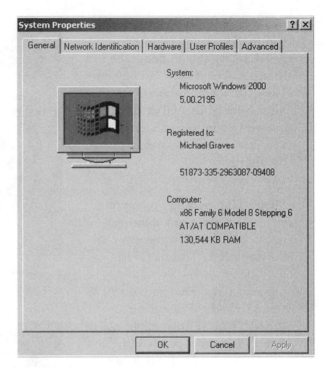

Figure L9.7   The System applet in the Win2K Control Panel

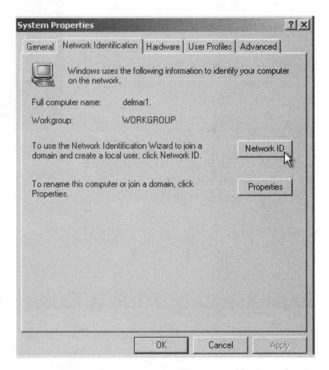

Figure L9.8   The Network Identification tab

6. Now take a look at that More... button (**Figure L9.10**). This button has one function: to change the primary DNS suffix for your computer. So what the heck is a primary DNS suffix and why would you want to change it? I'll answer the last question first. Unless an administrator tells you otherwise, you wouldn't want to change it. DNS is the Domain Name Services and is how different servers are named on the Internet or in a Win2K domain. The primary DNS suffix is a key part of your computer name. This feature only comes into play when the computer is a member of a domain. My only reason for showing you this screen is to point out that little check box labeled Change primary DNS suffix when domain membership changes. By default that box is checked. Now if I move this computer from the GRAVES.ORG domain to the DELMAR.COM domain, as soon as the administrator adds my computer to the DELMAR.COM domain, the computer automatically assumes the new suffix. 99.99% of the time, you don't want to fool with this setting. But if, for any reason, this box is deselected and you change domains, it can cause issues later on.

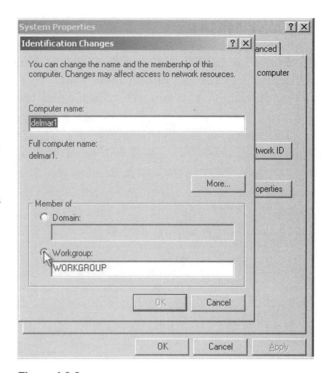

Figure L9.9
The Properties configuration screen

EXERCISE 2 REVIEW

1. What are three ways to get to Control Panel in Win2K?

2. Where does one change the computer's NETBIOS name in Win2K?

LAB REVIEW

1. What is the MMC, and what does it allow you to do?

2. Where would you go to change the NetBIOS name of a computer?

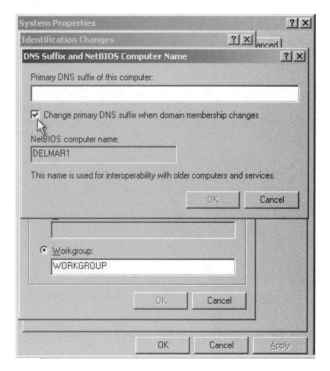

Figure L9.10   Changing the primary DNS suffix

3. You want to clear your history of recently opened documents. Where do you go to do that?

LAB SUMMARY

So Win2K isn't as different from Win9x as you initially thought, is it? Internally, the two OSs are as different as dogs and cats. But Microsoft wisely chose to keep the interface as similar as possible in order to ease the transition. As you will see in the next few labs, they weren't quite as friendly with WinXP.

INSTALLING WINXP

In this lab you will install Windows XP onto the lab machines. The only real difference between installing XP and Win2K is that there is no option for setting up a computer from floppy disks. If you have computers that do not boot to a CD-ROM, you must download a file called WinXP_EN_PRO_BF.EXE from Microsoft's website. As of this writing, that file was located at www.microsoft.com/downloads/details.aspx?displaylang=en&FamilyID=55820EDB-5039-4955-BCB7-4FED408EA73F. Since this is a copyrighted file, it is not possible for me to make it available to you from my website or delmar.com.

Since the procedure is identical to that of installing Win2K, I am not going to waste precious natural resources on 10 pages of identical text. The procedure is very automatic, but if you do require the assistance of instructions and screenshots, make use of Lab 8 in Part 2.

The materials you'll need for this lab include your student lab computer and a copy of Windows XP for each student.

The CompTIA objectives covered in this lab include the following:

1.2 Identify the names, locations, purposes, and contents of major system files.

2.1 Identify the procedures for installing Windows 9x/Me, Windows NT 4.0 Workstation, Windows 2000 Professional, and Windows XP, and bringing the operating system to a basic operational level.

2.3 Identify the basic system boot sequences and boot methods, including the steps to create an emergency boot disk with utilities installed for Windows 9x/Me, Windows NT 4.0 Workstation, Windows 2000 Professional, and Windows XP.

WORKING WITH ACCOUNTS

In the following exercises, you will be creating several different user and group accounts. These accounts will be used in later labs as you learn how to associate permissions and apply security to individual users and groups. You will also learn to copy an account, so that its permissions and settings are automatically applied to a new account that you've created. In the final exercise, you will rename and finally disable the account. In the exercises on groups, I have taken the liberty of using a Windows 2003 Server to show some advanced options. These are optional exercises. If you don't have access to either a Win2K Server or a Windows 2003 Server, simply follow along.

Sadly, this is an area that CompTIA does not deem sufficiently important to cover on the exam. It is, however, information critical to anyone working with computers on a professional level.

EXERCISE 1: CREATING A NEW ACCOUNT

Any time a new user is added to the network, that person will require a unique user account. This account, complete with user ID and password, is the ticket to the network. In order to create the account you will perform the following procedures.

1. Click Start→Programs→Administrative Tools→Active Directory Users and Computers (as shown in **Figure L11.1**).

2. You should get a screen similar to that in **Figure L11.2**. Highlight Users in the left pane then, in the right pane, right-click in any blank area. From the pop-up menu that appears, select New→User.

3. You should now have the screen shown in **Figure L11.3**. Type the user information as requested. In the User logon name field, type the User ID for that user. This must be a unique value. There can be no duplicate User IDs anywhere on the network. Click Next.

4. In the next screen (illustrated in **Figure L11.4**), you will be prompted to enter the user's password. You must enter it a second time in order to confirm the

Figure L11.1   A roadmap to Active Directory Users and Computers

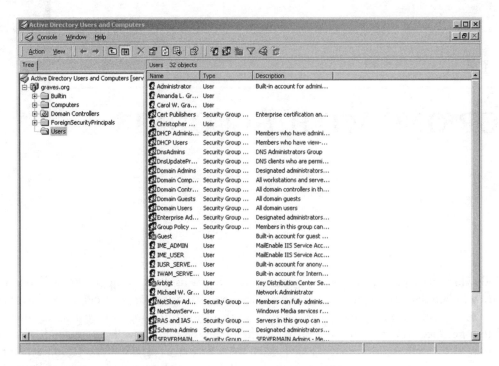

Figure L11.2   The Active Directory Users and Computers console

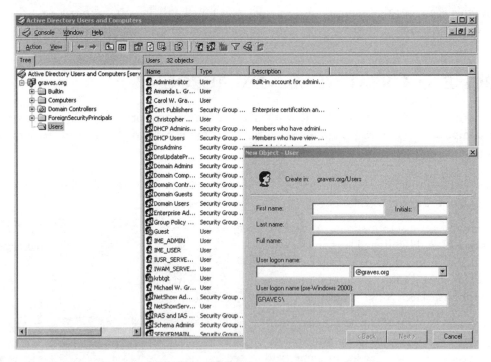

Figure L11.3   Fill in the user information and provide a unique User ID.

password. Should you inadvertently enter it differently the second time, the password will be rejected, and you will have to start again. There are four checkboxes beneath the password fields for password options:

a. User <u>m</u>ust change password at next logon: If this choice is selected, the first time users log onto their new accounts, they will be told that their password has expired and will be prompted to enter a new one (twice, for confirmation). This is the option to select if you want your users selecting their own passwords.

b. User cannot change password: As the phrase implies, after you have assigned a password, it is etched in stone. Only an account administrator or one with administrative privileges can change the password.

c. Password never expires: If this field is selected, the password will remain valid until changed by an account administrator or someone with administrative privileges. This is the case even if the password policy has been set to force users to change their password periodically.

d. Account is disabled: This option prevents anyone from logging onto the network using that particular account. It does not, however, delete any security settings or permissions.

5. For all accounts in your lab exercises, you will be using *password* as your standard password. This will prevent forgotten passwords from becoming an issue. Obviously, in a real-world scenario, this would be a very bad idea. Select User cannot change password and click Next.

6. You will get a summary screen like the one in **Figure L11.5.** All the information you typed will be displayed except for the password. Click Finish. Your new account has been established.

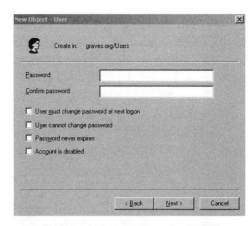

Figure L11.4 Next, enter a password, confirm it, and select the password options.

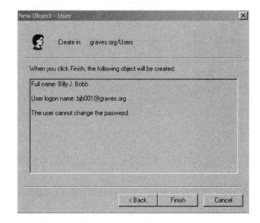

Figure L11.5 The User summary screen

7. Repeat the previous steps until you have created a total of twelve new accounts. Don't forget to use *password* as the password for all user accounts.

EXERCISE 1 REVIEW

1. Where do you go to create a new user account in WinXP?

2. What happens if you don't configure a password for XP?

EXERCISE 2: CREATING GROUP ACCOUNTS

Every network administrator quickly learns that managing groups of accounts all at once is much simpler than trying to manages the users one at a time. In this section, you will create two forms of group accounts. You will create local groups to manage resources, and you will create global groups to manage users. Later on you will use these global groups to manage permissions.

EXERCISE 2A: CREATING A LOCAL GROUP

1. Start Active Directory Users and Computers, just as you did in Exercise 1. Right-click a blank portion of the right pane and select New→Group from the pop-up menu. You'll get the screen shown in **Figure L11.6.**

2. For Group Name, type Documents. Notice that Windows fills in the field labeled Group Name (pre-Windows 2000) for you. Beneath those fields, on the left are the options D̲omain local and G̲lobal. Select D̲omain local. For now, don't worry about group type. Click OK.

3. Repeat the process, creating a group called DATA.

Figure L11.6   Creating a new group in Active Directory Users and Computers

EXERCISE 2B: CREATING A GLOBAL GROUP

1. Repeat the steps outlined in Exercise 2a. Name the group SALES. The one difference is that, instead of selecting D̲omain local in the second screen, select G̲lobal.

2. Repeat this process twice, creating global groups called MANAGEMENT and ADMINISTRATION.

EXERCISE 2 REVIEW

1. What is the difference between a local group and a global group?

2. What is the advantage to using groups?

EXERCISE 3: COPYING AN ACCOUNT

Now that you've created all those nearly identical groups and accounts the hard way, you'll learn how you can take an account that has been configured the way you want and make a new one using the first one as a template.

MANAGING GROUPS

When you're first getting started with this networking business, it sometimes gets confusing as to when to use groups and when not to. And when it comes to managing groups, what's with this *global group* versus *local group*?

It isn't really all that complicated. Local groups are used to manage local resources. You might have a database containing your customer information. In order to allow access to that database, you create a local group called DATA. Permissions are assigned at this level.

Global groups are used to give users with similar sets of responsibilities and resource needs the permissions they need to do their work. For example, you might have a global group called SALES. Every salesperson needs access to the same resources and generally needs the same permission sets. Therefore, when you hire a new salesperson, rather than go through the rigmarole of assigning those permissions independently (and remembering what they are), you simply create a new account and add it to the SALES group. In order to give all sales reps access to the database, you add the global group SALES to the local group DATA. In one step, all sales reps are given exactly the same permissions to use the database.

A simple little mnemonic will help you remember what's going on. AGLP. *A*ccounts go into *G*lobal groups, which are added to *L*ocal groups, which are given *P*ermissions.

1. The first thing you want to do is customize an account so that you know it is different than the others you created. Select one of the accounts that you created in Exercise 1 and double-click it in the right pane of Active Directory Users and Computers. You'll get the screen shown in **Figure L11.7.**

2. Click the tab labeled Member Of. Then click Add. This will give you the screen shown in **Figure L11.8.** As shown in the illustration, add this account to several different built-in groups. Also, add it to your newly created SALES group. Click OK.

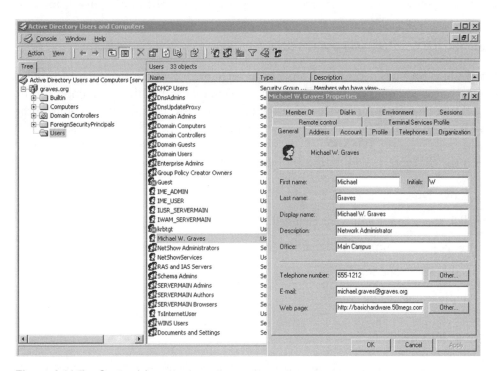

Figure L11.7 Customizing an account

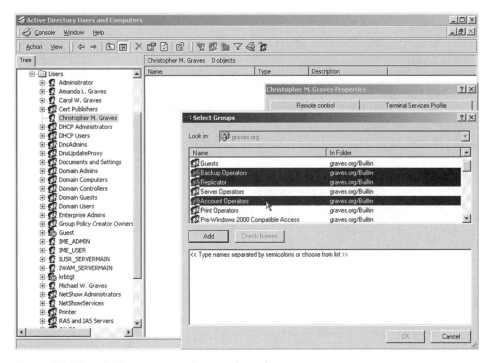

Figure L11.8 Making an account a member of a group

3. Open the Users Folder in the left pane of Active Directory Users and Computers and right-click the account you just modified. Select Copy. The screens that follow are the same ones you saw when creating a new account. That is because that's exactly what you're doing. Except this new account brings with it all the accoutrements of the account you copied.

Exercise 3 Review

1. Why would you want to copy an account as opposed to simply creating one from scratch?

Exercise 4: Renaming an Account

Once in a while it becomes necessary to rename an account. It's all too tempting to simply delete the old one and create a new one. However, there is a problem inherent in that procedure. It is not the user name, user ID, or password—or any combination of those—that identifies the account to the OS. The account is identified by a 32-bit number that was generated by the OS when the account was created. This is the account's security ID (SID). If you want to keep the entire history of the account intact, you need to keep the SID intact. You do this by renaming the existing account. Here's how to do it:

1. Open Active Directory Users and Computers and open the Users folder.

2. Right-click the account you want changed.

3. From the pop-up menu, select Rename.

4. Type in the new name for the account.

It's as simple as that.

Exercise 4 Review

1. What is the disadvantage of deleting the account of a user when he or she leaves the company?

2. What is the purpose of the SID?

Exercise 5: Disabling an Account

When a user leaves an organization, the first thing many administrators do is to delete that user's account. This can be (and frequently is) a critical error. The reason for not deleting a no-longer active account is the same one I gave for not creating a new account for an existing user. That SID is the ticket to the account's history. Should you need to access that account for any reason, it won't be possible if it was deleted. Simply recreating it won't work. Fortunately, disabling an account is one of the easiest things you'll ever have to do.

1. Open Active Directory Users and Computers.

2. Open the Users folder.

3. Right-click the account you want disabled.

4. Select Disable.

Later on, should you need to reactivate that account for any reason, you simply repeat the process, except that the option Enable will replace Disable. Select Enable, and the account will be reactivated.

LAB REVIEW

1. You have just created a new account for a user on the network, but you want him to select his own password. How do you make sure he has no choice but to create a new password?

2. What is the advantage of copying an account over creating a new one?

LAB SUMMARY

WinXP was designed from the ground up to be an OS that supports multiple users. Therefore, from an administrative standpoint, being able to manage those users is a critical ability. Although this lab only touches on managing accounts, without those accounts in place, the system is not much more secure than a Win9x box.

THE EVENT VIEWER

One of the better troubleshooting tools provided by Win2KS is one called Event Viewer, which collects information on different activities that are generated by either hardware or software action. These events range from benign to critical, and Event Viewer frequently can provide information that helps the administrator diagnose what led up to the event.

Event Viewer reports three degrees of severity in its logs.

Table 12.1 Event Viewer Severity Classifications

Symbol	Severity	Description
(i)	Information	Information describes the successful operation of an application, driver, or service.
(X)	Error	A significant problem, such as loss of data or loss of functionality.
(!)	Warning	An event that is not necessarily significant but may indicate a possible future problem.

With this information in mind, take a look at Event Viewer and see what you can find.

The only materials you'll need for this lab are your lab computers. The CompTIA objectives covered in this lab include the following:

1.5   Identify the major operating system utilities, their purpose, location, and available switches.

3.1   Recognize and interpret the meaning of common error codes and startup messages from the boot sequence and identify steps to correct the problems.

3.2   Recognize when to use common diagnostic utilities and tools. Given a diagnostic scenario involving one of these utilities or tools, select the appropriate steps needed to resolve the problem.

EXERCISE 1: AN OVERVIEW OF EVENT VIEWER

1. Click Start→Programs→Administrative Tools→Event Viewer. The screen in **Figure L12.1** will appear. Here, I added a computer that had not been on my network since rebuilding the domain. That gave me some very interesting errors to examine.

2. Notice in the left-hand pane that there are six different aspects of system performance that are logged.

    a. Application Log: Here is where events logged by applications or programs are recorded.

    b. Security Log: Events such as invalid logon attempts are recorded here. If you have auditing enabled, this is where auditing events will be stored.

    c. System Log: Any event generated by a system component, whether it be hardware (such as a memory error) or OS related, will be stored here.

    d. Directory Service Log: Events directly related to Active Directory are recorded here.

    e. DNS Server: Events generated by DNS services are stored here.

    f. File Replication Service: As its name implies, events generated by the File Replication service are stored here. An example of this event would be the failure of two domain controllers to successfully replicate the SID or SYSVOL.

3. Highlight the Application Log in the left-hand pane. Click Action. Click Properties. This will bring up the screen in **Figure L12.2**. The two tabs in this screen are General and Filter. Under General, you can configure several different things.

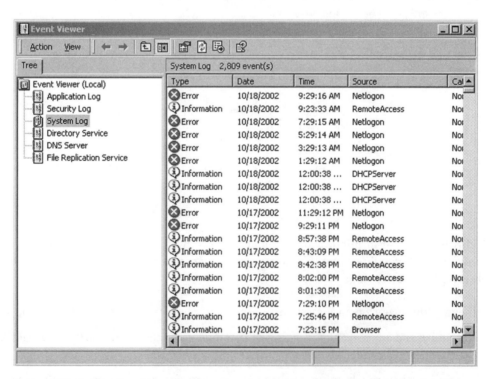

Figure L12.1   The Event Viewer screen

4. Under Display Name, change Application Log to Server Applications. Click Apply and then OK. Notice the change on your Event Viewer Screen.

5. Go back to the Properties screen for the Application Log and change it back to its original name. Notice that the default <u>M</u>aximum log size is 512KB. Also, by default, when the log size exceeds 512KB, it is set to O<u>v</u>erwrite events older than seven days. If hard drive space is not an issue, I suggest that you increase your log size to 2048KB (2MB) and that the selection <u>O</u>verwrite events as needed be selected. Make these changes

6. Another option on this screen is to <u>C</u>lear log. Clicking this button will completely delete all events recorded in this log. You will be asked whether you want to save the log files before you continue (**Figure L12.3**). You don't really want to clear your log, so click Cancel and move on.

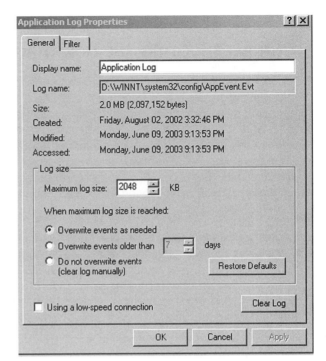

Figure L12.2   The Application Log Properties window

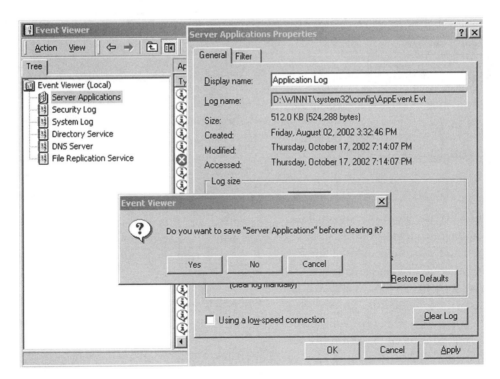

Figure L12.3

TROUBLESHOOTING THE NOS THROUGH EVENTS

When you're first getting starting with this network administration thing, it can appear to be overwhelming at first. You'll soon get over that. You'll find that all providers of network operating systems provide a substantial amount of support for their products. In Windows 2000 and XP, the Event Viewer provides information on what caused a failure. If this doesn't help, you can take it a step farther and make use of their TechNet services on Microsoft's web page (currently at www.microsoft.com/technet). A search of key words from the message is very likely to bring up several articles related to your problem. Novell offers very similar services, and Linux help can be obtained from Red Hat, Mandrake, and numerous other Linux vendors.

Another good resource for Microsoft users is the Windows 2000 or Windows XP Resource Guide. Again, Novell provides similar references for its NOS. In this kit you will find reams of information. Nearly every error message generated by the NOS is explained, and most causes of service or driver failure can be found in this guide. These are not inexpensive books, but compared to the cost of your NOS and/or the cost of the administrator's time, the resource guide is an essential tool for any administrator.

SUMMARY OF EXERCISE 1

1. Where would you find the Event Viewer in Windows 2000?

2. If a service fails to start, what kind of icon will it display?

3. You have configured your server to audit failed logon attempts. Where would reports of these events be logged?

EXERCISE 2: ANALYZING AN EVENT

1. For this exercise, go to the System Log. This is where you will find the most events. Since it is impossible for me to predict what all of your systems are going to look like, follow closely with the illustrations in addition to performing these steps on your own computer. My descriptions of events and other information will be based entirely on the illustrations.

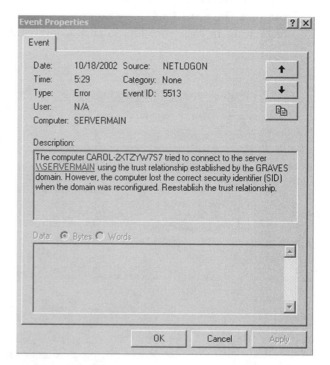

Figure L12.4 The Event Properties window

2. Double-click any error your System Log may be reporting. If there are no errors, find an Information event. Lacking that, double-click any event but follow the text carefully. Double-clicking any event will bring up the Event Properties screen (**Figure L12.4**).

3. If you are looking at an error message, such as the one in the illustration, the description screen will tell you precisely what failed. In the case of the error in the illustration, you are even told how to fix the problem. All I need to do is rejoin this computer to the domain.

4. Close Event Viewer.

SUMMARY OF EXERCISE 2

1. What kinds of information can you learn from an event if you double-click it?

2. If a service fails to start, what can you learn by examining the event?

LAB REVIEW

1. What are the different logs kept by Event Viewer?

2. Describe the three little icons used by Event Viewer to identify the severity of an event and explain what each one means.

LAB SUMMARY

The Event Viewer is one of the first places a systems administrator goes for information when something goes wrong. This lab provided a glimpse into the reasoning for that. The Event Viewer is actually a very powerful tool when you get the hang of using it, you'll make it one of your first stops as well.

BACKUP AND RECOVERY

In the following exercises, you're going to get a brief overview of the Windows Backup utility. Then you'll go through the process of performing a backup and subsequently deleting and restoring the data you backed up. Optimally, in order to do this lab, each computer should be equipped with a tape backup unit. However, it is assumed that is not the case in the majority of classrooms, and the backup will be done to file in these exercises.

For this lab, you'll need your student computers, tape drive (if possible), and backup tapes. There's only one CompTIA objective covered in this lab:

1.2 Identify basic concepts and procedures for creating, viewing, and managing files, directories, and disks. This includes procedures for changing file attributes and the ramifications of those changes (for example, security issues).

EXERCISE 1: AN OVERVIEW OF THE BACKUP UTILITY

1. To start Win2K backup, click Start→Programs→Accessories→System Tools→Backup. This will bring up the screen shown in **Figure L13.1**.

2. The three options shown in this window are Backup Wizard, Restore Wizard, and Emergency Repair Disk. You won't be working with any of the Wizards in this lab. You'll be learning to perform manual backups and recoveries.

3. Click the Backup tab. You should now have a window like that in **Figure L13.2**. Note that you can back up any or all of your local drives, from your CD-ROM drive, the System State, and from Network Places. If you click your C: (or any other) drive, you will see the various folders on that drive. You can choose which files and folders to back up.

4. At the bottom of the screen you have two other options. Backup destination allows you to select where that file is going to be stored. If you have no tape drive installed on your machine,

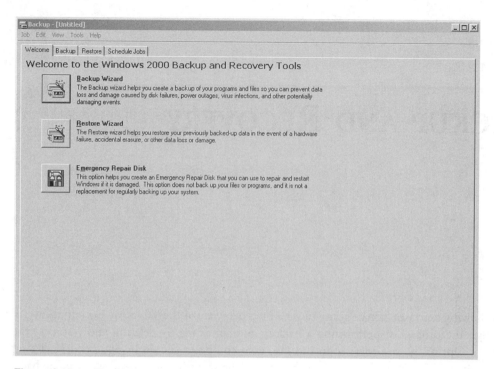

Figure L13.1   The Windows Backup utility

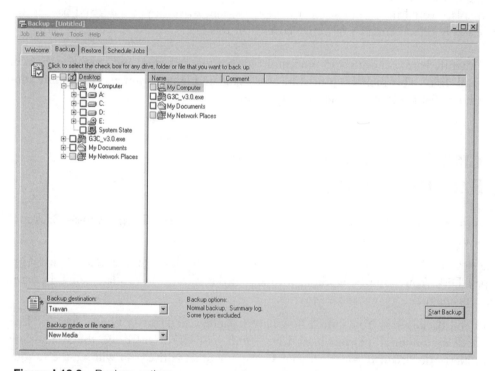

Figure L13.2   Backup options

the only option will be File. Note in **Figure L13.3** that I have the option of backing up to a miniQIC, Travan, or File. Backup <u>m</u>edia or file name allows you to indicate the form of medium you're going to use or select a file name (with full directory path) for your backup.

5.   Now click the Tools option in the menu and select Options (**Figure L13.4**). Under Backup Type you can select Normal, Copy, Differential, Incremental, or Daily. You'll be looking at this section in closer detail in Exercise 2, so for now, move on to the next section.

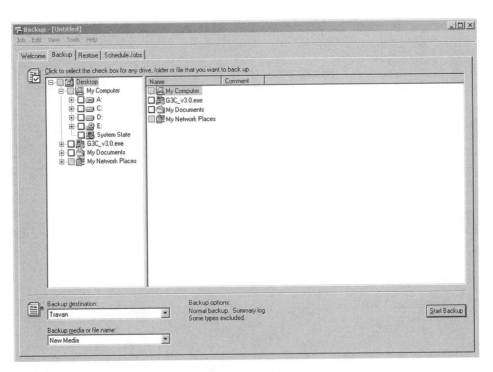

Figure L13.3 Selecting the destination for your backup

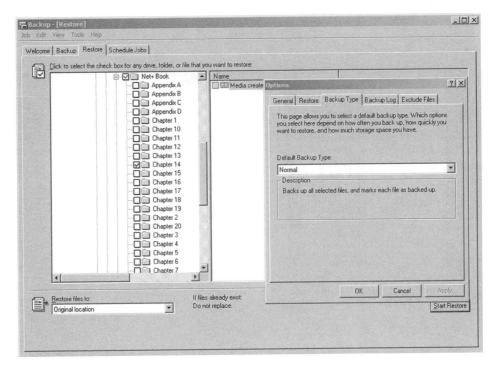

Figure L13.4 Selecting the backup type

6. Click the Restore tab. In this screen (**Figure L13.5**), you are presented with all possible locations on your computer where a backup could exist. If you have no tape drive installed on your machine, the only option available to you will be File.

7. In Figure L13.5, I've opened the Travan option to show that my last backup included my C: drive and my D: drive. In **Figure L13.6**, I've opened the contents of the D: drive. When you do this, your tape drive will go into action as it tries to load the contents of that folder. When

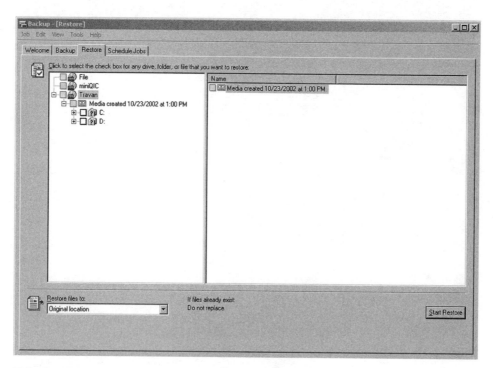

Figure L13.5 Restore options in the Windows Backup utility

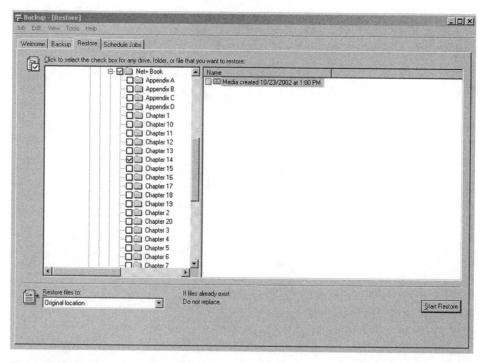

Figure L13.6 The Windows Backup utility allows you to selectively restore files.

it has done that, it will rewind the tape. This can take several minutes on Travan drives. DAT or DLT drives are usually somewhat quicker.

8. Notice that by clicking just a single subdirectory or file (as I've done in Figure L13.6), I can restore just that file or directory. By clicking an entire drive, I will restore the contents of that entire drive.

SUMMARY OF EXERCISE 1

1. Where do you find the backup utility in Windows XP?

2. What is the difference between a differential backup and an incremental backup?

3. What is the difference between copying your files and backing them up?

EXERCISE 2: PERFORMING A BACKUP

In this exercise, you will back up a single directory on your hard drive to a file. In order to expedite the procedure, you will select a small directory. You will use the one you created using your own first name.

1. On the Backup screen, click the Backup tab. Highlight the hard disk drive that contains your folder, and then click the checkbox next to that folder (**Figure L13.7**).

2. Under Backup destination, select File, and under Backup media or file name, change A:\backup.bkf to C:\backup.bkf. (Note that in the real world, backing up files from your hard drive to your hard drive is not a very sane practice. If your hard drive fails, it all fails, not just selected directories!)

> **NOTE:** Most backup software still offers the option of backing up your files to floppy. This is a viable option for backing up just a few files, but it can also be used when you have a file that is too large to fit onto a floppy disk. The Backup utility will split large files onto several floppy diskettes. If you want, you can back up your entire hard drive to a collection of floppy diskettes. However, I would like to go on record as saying that the idea of backing up my 40GB hard disk to 27,778 diskettes is not a project that is close to my heart.

3. Click Start Backup. Since you've selected an extremely small backup set, this should take only a second or two. When it is finished you will get a screen like the one in **Figure L13.8**.

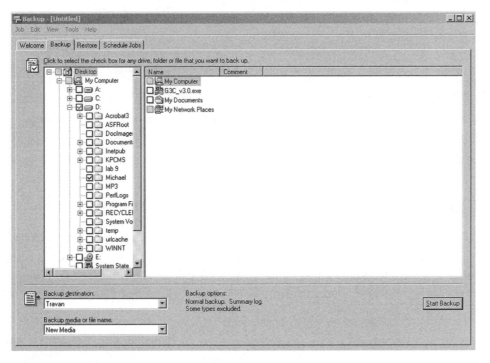

Figure L13.7   Selecting the files to be backed up

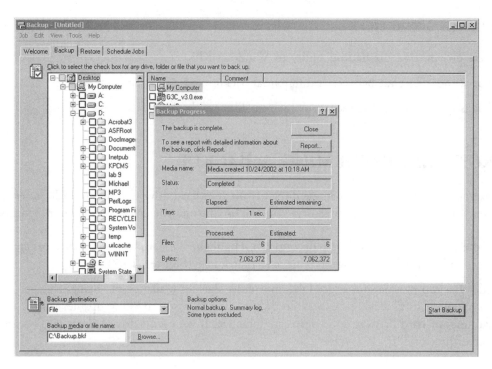

Figure L13.8   The Backup Progress screen

4. Here you are informed of the time and date of the backup along with whether it was successful or not. In addition, you can see how long the backup took, how many files were processed, and the total number of bytes that were backed up. Now click the Report button.

5. For **Figure L13.9**, I performed a backup to my tape drive of the My Documents folder on my hard drive. In order to generate some failure messages, I left Microsoft Word, along with several documents, open. Notice the number of files that were skipped because they were in use. This is a key reason why network backups should be performed at a time when the fewest users (none at all, if possible) will be active on the network.

SUMMARY OF EXERCISE 2

1. What types of destination locations are supported by Windows XP?

2. Why wouldn't you want to back up your hard drive to floppy diskettes?

3. How can you find out whether all of your files were backed up when the operation is completed?

4. If not all files were backed up, what are some possible causes?

EXERCISE 3: PERFORMING A RESTORE OPERATION

Restoring data to a hard disk drive need not be cause for panic. From your restore file, you have the option of performing a complete restore (as would be required after a hard disk failure once the disk was replaced) or restoring selected files (as might be necessary if a single file is inadvertently deleted, overwritten, or corrupted). In this exercise, you will delete the folder that you just backed up and use your backup file to recover the lost data.

1. First of all, go into Windows Explorer and delete your folder. Since you shared this folder out in an earlier lab, you will be warned that others might be using the folder. Click OK.

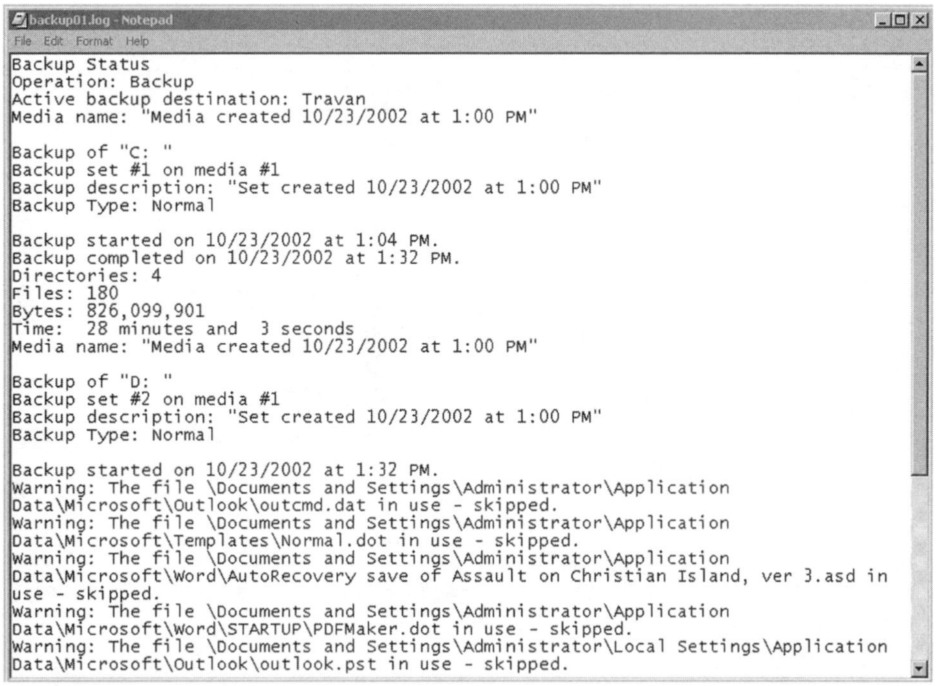

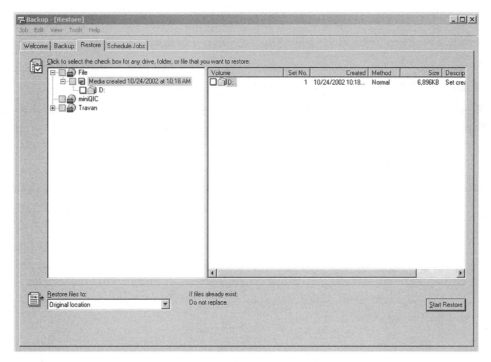

Figure L13.9 Viewing the Backup log

Figure L13.10 Selecting the files to be restored

2. In the Backup utility, click the Restore tab. Click the + next to File and highlight the media set you created in Exercise 2 (**Figure L13.10**).

3. Check the box next to your folder. Make sure that the option Restore files to has Original location selected, and click the Start Restore button. The Confirm Restore screen will appear (**Figure L13.11**) and offer the choice of starting your Restore or selecting Advanced Options.

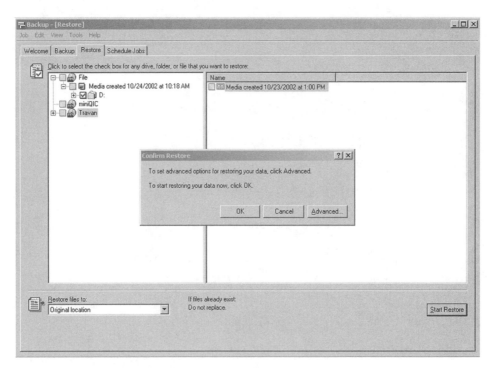

Figure L13.11 The Confirm Restore screen

4. Click the Advanced button. You don't really need to use any of these options, but now is a good time to explore them, since you're here anyway. These options include:

   a. Restore security: This should be checked by default. It makes sure that all permissions assigned to this folder and the files it contains remain intact.

   b. Restore Removable Storage database: Removable Storage is a Windows service that allows applications to access and share resources stored on removable media. Unless you've installed and configured Removable Storage on your system it is not necessary to select this option.

   c. Restore junction points, and restore file and folder data under junction points to the original location: A junction point is a physical location on your hard drive that points to another physical location or another storage device. It's a good idea to always leave this box checked. There may be no junction points required, but it's better to have it and not need it than to need it and not have it.

   d. When restoring replicated data sets, mark the restored data as the primary data for all replicas: This is most likely grayed out on your screen. It ensures that information used by the File Replication service knows whether or not this data should be replicated to other servers on the network.

   e. Preserve existing volume mount points: This particular option really only makes a difference when restoring an entire drive. If you are installing a new drive and it has been partitioned, it is best if this option were not checked. Otherwise, new partitions will be created on the drive.

5. For the purposes of this exercise, none of these options are needed. So click Cancel to get back to the Confirm Restore screen and click OK. You'll be prompted to enter the location of the backup file. C:\backup.bkf should be the default location (**Figure L13.12**). Click OK.

6. You will briefly seen a Restore Progress screen flashing the files as they are restored, and then Backup will settle into the screen shown in **Figure L13.13**. As you can see, it is identical to the

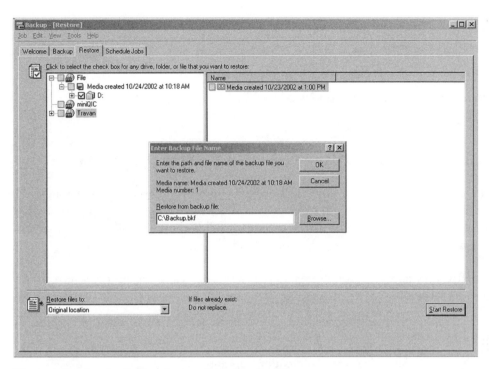

Figure L13.12   Confirming the location of backup files

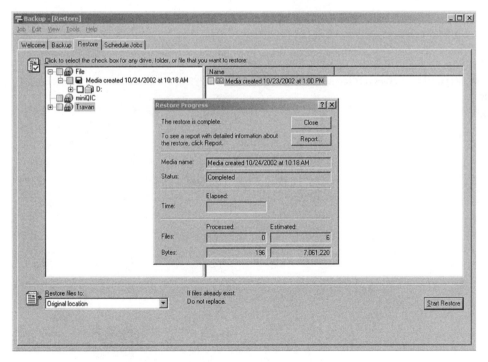

Figure L13.13   The Restore Progress screen

Backup Progress screen you saw in the previous exercise. It, too, offers the option of viewing a report, which will also be identical to the one you looked at in that exercise.

7.   Go to Windows Explorer once again. You should be able to browse to your folder and see that all contents are intact.

Summary of Exercise 3

1. One of your users has inadvertently overwritten a critical file on the server. Can you get just that file back, or do you have to restore the whole system?

2. Name some of the advanced options available.

Exercise 4: Scheduling Unattended Backups

In this area, the Windows Backup utility is a substantial improvement over previous versions of Microsoft's backup utilities. In this exercise, you'll see how you can schedule different backups to occur on different days of the week. You'll set up your machines to do a full backup on Friday evening at 8:00 P.M. and then do incremental backups at the same time Monday through Thursday.

1. In the Backup utility, click the Schedule Jobs tab. This will bring up the screen shown in **Figure L13.14**.

2. Click <u>A</u>dd Job in the lower right-hand corner of the screen. This starts the Backup Wizard (**Figure L13.15**).

3. Click <u>N</u>ext. The following screen offers three options.

    a. Back up <u>e</u>verything on my computer: Does just what it suggests. However, don't put too much faith in what it says. As you saw earlier, open or locked files will not be backed up.

    b. Back up selected <u>f</u>iles, drives, or network data: Allows you to pick and choose what material should be backed up.

    c. Only back up the <u>S</u>ystem State data: This information includes the files that make up the registry, the COM+ registration database, and all system boot files.

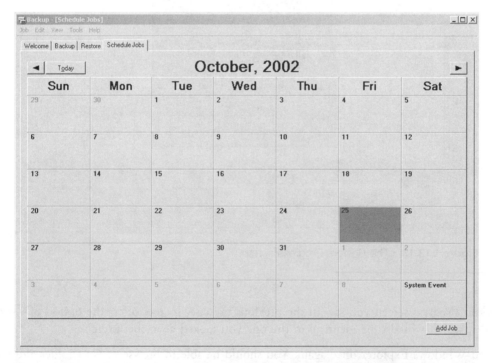

Figure L13.14   The Windows Backup utility allows you to schedule unattended backups.

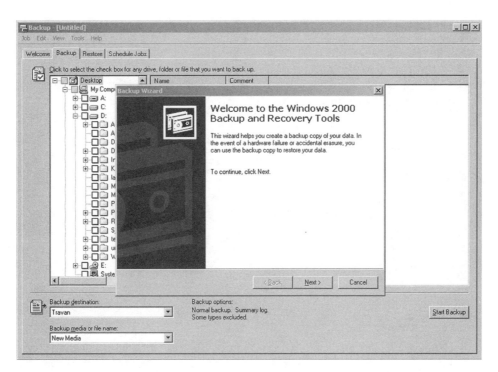

Figure L13.15   Adding a new job to the Backup schedule

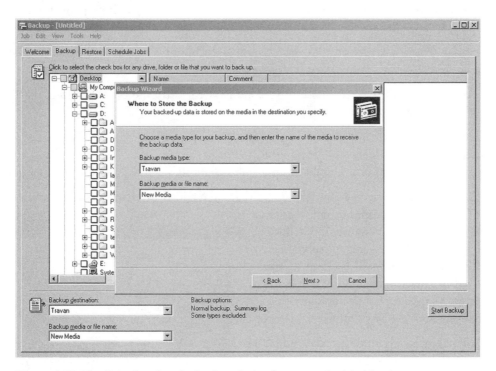

Figure L13.16   Selecting the destination device for your scheduled backups

4.   Select Back up everything on my computer and click Next. This brings up the screen in **Figure L13.16**, where you are prompted to select the destination for your backup. In my illustration, I'll select my Travan drive. Depending on your setup, you should either select the tape drive that is available, or select File.

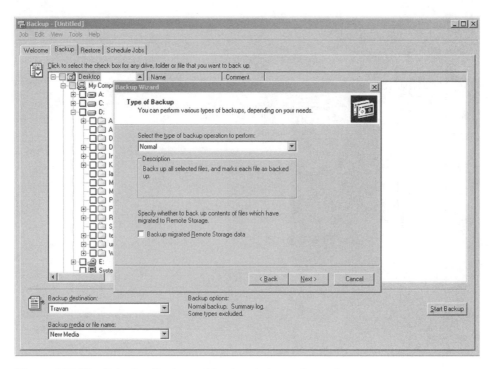

Figure L13.17   Selecting the type of backup to be performed

5. In **Figure L13.17**, you're prompted to enter the type of backup you'll be performing. The options are

    a. Normal: Copies all selected files and clears the attribute bit, marking them as backed up. If an entire drive was selected, this is the equivalent of a Full backup.

    b. Copy: It copies all selected files but does *not* clear the attribute bit. Therefore, the files will not be marked as backed up.

    c. Incremental: Copies any files that were added or changed since the last Normal or Incremental backup. It clears the attribute bit, marking the files as backed up. All subsequent incremental backups will now back up all files changed since the last incremental backup.

    d. Differential: This selection copies all files that were added or changed since the last Normal or Incremental backup, but does *not* clear the attribute bit. Therefore, files will not be marked as backed up. All subsequent differential backups will back up all files added or changed since the last Normal or Incremental backup, but not those that changed since the last Differential backup.

    e. Daily: Copies only files that were added or changed on the day the backup is created. The attribute bit is not cleared.

6. Since you are creating a full backup, select Normal and click <u>N</u>ext. The Backup Wizard now asks you how you want to back your data up (**Figure L13.18**). Two independent options are offered.

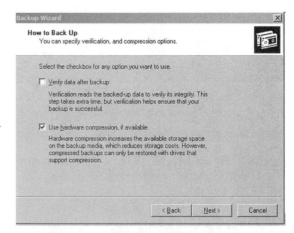

Figure L13.18   Configuring how the backup is to be done

a. Verify data after backup: This option compares each copy of the file to the original when the backup has been completed. This can add a substantial amount of time to the backup but adds security for your data.

b. Use hardware compression, if available: This allows you to pack more data onto a single tape and reduces the amount of time it takes for a backup to be completed.

7. Click Next. You'll be given the Media Options screen (**Figure L13.19**). Normally, you would be using the same tapes over and over again. In that case, you would make sure that the option Replace the data on the media with this backup is selected. If the data is sensitive data, you can add a bit more security by selecting the option Allow only the owner and the Administrator access to the data and to any backup appended to this media. The latter option is not selected by default.

8. The Backup Label screen now appears (**Figure L13.20**). This is the information that you should write onto the label of the tape before storing it. Click Next.

9. Now you'll be prompted to establish when you want your backup to occur (**Figure L13.21**). Select Later and give your job a name. Call it Weekly Full.

10. Now click the Set Schedule button. Under Schedule Task, select Weekly. Under Start Time, select 12:00 A.M. Under Schedule Task Weekly select 1 for Every and click the Friday checkbox, as in **Figure L13.22**. Make sure that all other days are deselected.

11. Click OK and on the Backup Wizard screen click Next. This will bring up the Completing the Backup Wizard screen as shown in **Figure L13.23**. Click Finish and you're done.

12. Now, on the Schedule Jobs calendar, little icons with a blue N (for Normal) will appear on all Fridays from this date forward.

13. To schedule daily differential backups to occur, repeat the preceding procedure except for two key differences. In step 5, where you configured backup type, select differential. In

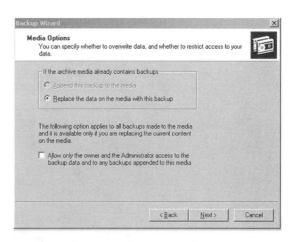

Figure L13.19   You have to tell the Backup utility how media is to be handled.

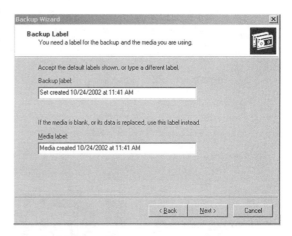

Figure L13.20   The Windows Backup utility is even gracious enough to remind you to label your backup tapes.

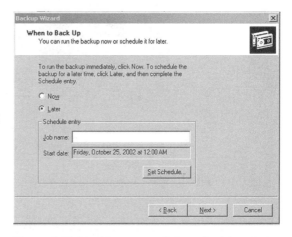

Figure L13.21   You must tell the scheduler what time you want these backups to occur. Choose a time when there will be the fewest users on the network.

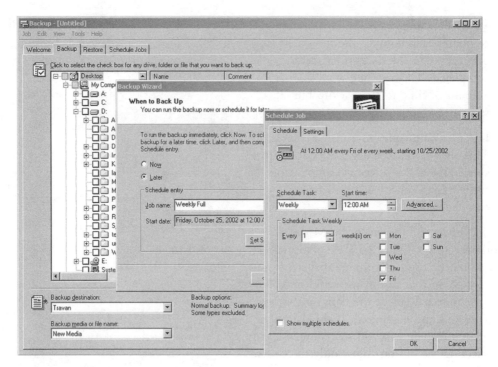

Figure L13.22 Scheduling daily or weekly events

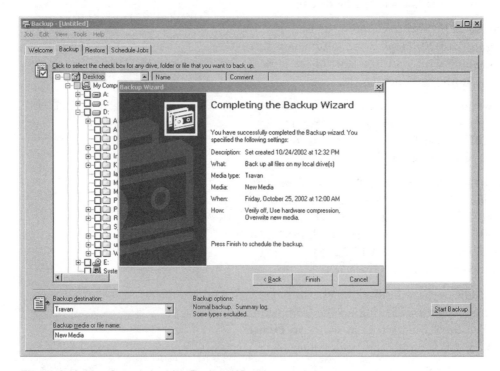

Figure L13.23 Completing the Backup Wizard

step 10, where the events are scheduled, select Weekly as before, but check the boxes for Mon, Tue, Wed, and Thu. Under Job name call it Daily Differential. Finish the Wizard, and now icons will appear on all Mondays through Thursdays with a green D (for differential). Your calendar should now resemble **Figure L13.24**.

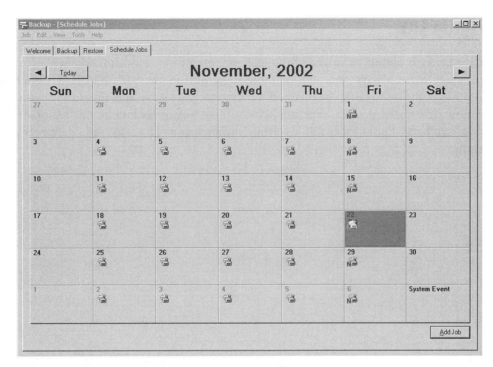

Figure L13.24 A Scheduling Calendar with events in place

14. Now the most important step of all. Make sure there is a tape in the drive for each day an event is scheduled. It may seem like a no-brainer, but an empty tape drive is undoubtedly the most common cause of backup failure there is.

> NOTE: You may notice if you look carefully that there is no option in the Backup Scheduler to delete a scheduled event. If you need to delete a scheduled backup for any reason, open Control Panel and double-click Scheduled Tasks. Highlight the job you want to blow away and press the <delete> key on your keyboard. It'll ask you whether you're sure. Click OK and the jobs are gone.

SUMMARY OF EXERCISE 4

1. You want to schedule your server to perform a Full backup every week, starting at 12:00 A.M. on Saturday. But you want your daily backups to be differential backups and to run at 10:00 P.M. every other night. How do you keep these from conflicting?

2. What is the primary cause of backup failure in unattended scheduled backups?

3. You need to delete one or more scheduled backup jobs, but there is no delete function in the Backup utility. How do you delete them?

LAB REVIEW

1. You want to set up a backup routine in which you do a full backup once a week. Then every night for the rest of the week you want to back up only the data that was added or changed that day. Which method do you select for those subsequent days?

2. You have a scheduled backup that you want to remove. Where do you go to do this?

LAB SUMMARY

Backup and recovery is undoubtedly one of the most important jobs a systems administrator does. Yet, in my experience, it has always been one of those tasks put on the back burner. I have worked in organizations where the information maintained was extremely sensitive and vital to the existence of the place. And more than once I saw my "superiors" blow off the nightly backup for one reason or the other. Don't get into that habit. Put together a solid scheme and stick to it. That way, should disaster ever occur, you'll be ready.

ADDITIONAL RESOURCES

Throughout this manual, I've referred to several different resources on the Web. These can be broken down into three categories, as far as these labs are concerned. Acquisitions resources are the various companies that provide parts and/or services that you'll need as a technician. Reference sites are good for researching certain topics, and finally none of us can exist without a little technical support now and then. The following section is broken down into those three sections, providing Web addresses, physical addresses, and telephone numbers where appropriate. Note that many Web resources don't provide physical information, so neither did I.

ACQUISITION RESOURCES

You can probably purchase troubleshooting supplies, components, and replacement parts in a couple of million different places. I have neither the time, the space, nor the inclination to list them all. Therefore, I'll list only the ones I mentioned in the textbook or in the lab manual.

Ultra-X, Inc.
www.uxd.com
Tel: 408-261-7090
Fax: 408-261-7077

JDR Microdevices
www.jdr.com
1850 South 10th Street
San Jose, California 95112-4108
Tel: 1-800-538-5000
Fax: 1-800-538-5005

Tiger Direct
www.tigerdirect.com
7795 W. Flagler St. Suite 35
Miami, Florida 33144
Tel: 1-800-800-8300
Fax: 305-415-2202

Ebay
www.ebay.com

Insight Components
www.insightcomponents.com
Tel: 800-723-8282

TuffTEST
www.tufftest.com
130 Alto Street
San Rafael, California 94901
Tel: 1-415-456-2200
Fax: 1-415-456-2244

REFERENCE

Most major manufacturers provide excellent collections of white papers describing the different technologies that go into their products. In the course of my writing, I draw heavily on these resources. After all, who knows better about how a particular product works than the people who make that product? In addition to the manufacturers, a number of people and organizations have taken the time and effort to share their knowledge on the Web. I've included a few resources that I think you'll find useful. Please don't consider this to be an all-inclusive list.

International Business Machines, Inc.
www.ibm.com
1133 Westchester Avenue
White Plains, New York 10604

Dell Computer Corporation
www.dell4me.com
One Dell Way
Round Rock, Texas 78682
Tel: 1-888-560-8324

Seagate Technology, Inc.
www.seagate.com
920 Disc Drive
Scotts Valley, California 95066
Tel: 1-831-438-6550

Micron Technology, Inc.
www.micron.com
8000 South Federal Way
Post Office Box 6
Boise, Idaho 83707-0006
Tel: 208-368-4000
Fax: 208-368-2536

Hardware Central
www.hardwarecentral.com

The PC Guide
www.pcguide.com

Basic Hardware
www.mwgraves.com

Intel Corporation
www.intel.com
2200 Mission College Blvd.
Santa Clara, California 95052
Tel: 800-628-8686

AMD Corporation
One AMD Place
P.O. Box 3453
Sunnyvale, California 94088
Tel: 408-749-4000

TECHNICAL SUPPORT

Most of the manufacturers of computers and computer components provide extensive technical support for their products, both online and by telephone. However, you might find a number of additional sites useful in terms of support issues and drives. A few of these are listed in this section. These are listed by Website only. Websites come and websites go, so don't be sending me hostile letters if you try to look up a site, and it's no longer there.

www.bootdisk.com

www.mrbios.com

www.freewarehome.com

download.com (CNET)

www.windrivers.com

www.driverguide.com

www.motherboards.org

Notes:

NOTES: